Mathematical Sciences Research Institute
Publications

6

Mathematical Sciences Research Institute Publications

1 Free/Uhlenbeck: Instantons and Four-Manifolds

2 Chern: Seminar on Non Linear Partial Differential Equations

3 Lepowsky/Mandelstam/Singer: Vertex Operators in Mathematics and Physics

4 Kac (editor): Infinite Dimension Groups with Applications

5 Blackadar: K-Theory for Operator Algebras

6 Moore: Group Representations, Ergodic Theory, Operator Algebras, and Mathematical Physics

Forthcoming

Chern/Griffiths: Exterior Differential Forms
Gersten: Essays in Group Theory

Group Representations, Ergodic Theory, Operator Algebras, and Mathematical Physics

Proceedings of a Conference in Honor of George W. Mackey

Edited by C.C. Moore

Springer-Verlag
New York Berlin Heidelberg London Paris Tokyo

C.C. Moore
Department of Mathematics
University of California at Berkeley
Berkeley, CA 94720

Mathematical Sciences Research Institute
1000 Centennial Drive
Berkeley, CA 94720

AMS Subject Classification: 22E45, E46, E47

Library of Congress Cataloging-in-Publication Data
Group representations, ergodic theory, operator
algebras, and mathematical physics.
(Mathematical Sciences Research Institute publications ; 6)
Papers presented at a conference sponsored by
the Mathematical Sciences Research Institute and
held May 21–23, 1984 in honor of Professor
George W. Mackey.
Includes bibliographies.
1. Lie groups—Congresses. 2. Representations
of groups—Congresses. 3. Ergodic theory—
Congresses. 4. Operator algebras—Congresses.
5. Mathematical physics—Congresses. 6. Mackey,
George Whitelaw, 1916– . I. Moore, C. C.
(Calvin C.) II. Mackey, George Whitelaw, 1916–
III. Mathematical Sciences Research Institute
(Berkeley, Calif.) IV. Series.
QA387.G75 1987 512′.55 86-29733

Printed and bound by R.R. Donnelley & Sons Harrisonburg, Virginia.
Printed in the United States of America.

9 8 7 6 5 4 3 2 1

ISBN 0-387-96471-1 Springer-Verlag New York Berlin Heidelberg
ISBN 3-540-96471-1 Springer-Verlag Berlin Heidelberg New York

George W. Mackey

Editor's Preface

The Mathematical Sciences Research Institute sponsored a three day conference, May 21-23, 1984 to honor Professor George W. Mackey. The title of the conference, Group Representations, Ergodic Theory, Operator Algebras, and Mathematical Physics, reflects the wide ranging interests in science that have characterized Professor Mackey's work. The conference provided an opportunity for his students, friends and colleagues to honor him and his contributions. The conference was attended by over one hundred people and the participants included five mathematical generations -- Professor Mackey's mathematical father, Marshall Stone, many mathematical children, grandchildren, and at least one mathematical great-grandchild.

This volume is a compendium of the scientific papers presented at the conference plus some additional papers contributed after the conference. The far ranging scope of the various articles is a further indication of the large number of fields that have been affected by Professor Mackey's work.

Calvin C. Moore
Berkeley, CA
Feb, 1986

Table of Contents

AMBIGUITY FUNCTIONS AND GROUP REPRESENTATIONS

L. Auslander[1] and R. Tolimieri

Dedicated to George, W. Mackey

0. INTRODUCTION

P.M. Woodward popularized the concept of ambiguity functions in his book ↑Wo←. Ambiguity functions did not yield to the standard theory of Abelian harmonic analysis and though studied by numerical methods their theory has not advanced much since the fundamental work of C. Wilcox ↑W←. In recent years, the Wigner transform, a close relative of the ambiguity function, has gained a lot of attention in engineering circles that are concerned with speech and its synthesis and non-intrusive methods for determining fluid flow in the human body. There seem to be three major theoretical problems that are of great interest in this subject:

(1) Find all ambiguity functions;

(2) Develop a theory for sampling ambiguity functions;

(3) Complete Woodward's program of giving an information theoretic model that includes the ambiguity function.

In this paper we will describe how representation theory of nilpotent groups begins to give information about (1) and (2).

1) Received partial support from the NSF during preparation of this paper.

1. SIMPLIFIED DISCUSSION OF THE USE OF AMBIGUITY FUNCTIONS IN RADAR

We will assume one target and see how to determine the distance of the target from the origin, r(t), and $\frac{dr}{dt}$ = v(t), abbreviated as velocity. Our problem can then be formulated as follows:

<u>Problem</u>: By transmitting a wave or pulse s(t) from the origin for $-T \leqslant t \leqslant T$ and receiving an echo e(t) determine r(0) and v(0).

The first step is to replace the pulse s(t) with its waveform $u_s(t)$. Before doing this, let us introduce the following notation. Consider $\mathbb{R}$ with parameter t, (t can be considered as time) and its dual group $\hat{\mathbb{R}}$ with parameter f (f can be considered as frequency). For $g(t) \in L^2(\mathbb{R})$ let its Fourier transform in $L^2(\hat{\mathbb{R}})$ be denoted by $\hat{g}(f)$.

Because s(t) is real valued $\hat{s}(-f) = (\hat{s}(f))^-$ and so s(t) is determined by its positive spectrum. Define

$$\psi_s(t) = \int_0^\infty \hat{s}(f)e^{2\pi ift}df$$

Then

$$\psi_s(t) = s(t) + i\sigma(t)$$

where $\sigma(t)$ is the Hilbert transform of s(t), or using principal part integrals

$$\sigma(t) = \frac{1}{\pi}\int_{-\infty}^{\infty} \frac{s(\tau)}{t-\tau}\, d\tau.$$

Notice that $\|\psi_s(t)\|^2 = 2\|s(t)\|^2$, where $\| \ \|$ denotes L^2 norm. There are three assumptions that are usually made, and we will assume them for the rest of this section.

(1) $\|\psi_s(t)\|^2 = 1$

(2) $t_0 = \int^{\mathbb{R}} t\,|\psi_s(t)|^2 dt < \infty$

(3) $f_0 = \int f\,|\hat{\psi}_s(f)|^2 dt < \infty.$

Definition. The waveform $u_s(t)$ of the pulse $s(t)$ is defined by

$$u_s(t) = \psi_s(t+t_0)e^{-2\pi i f_0(t+t_0)}$$

Notice: (1) $\|u_s(t)\|^2 = \|\psi_s(t)\|^2 = 1$

(2) $s(t) = \mathrm{Re}\{u_s(t-t_0)e^{2\pi i f_0 t}\}$

For one target and under certain physical assumptions, the echo is given by

$$e(t) = \mathrm{Re}\{\psi_e(t)\}$$

where $\psi_e(t) = \psi_s(t-x_0)e^{-2\pi i y_0 t}$ and $x_0 = \frac{2}{c}r(0)$, $y_0 = \frac{2}{c}f_0 v(0)$, c = speed of light. If we could determine x_0, y_0 we would be done. In order to do this, define

$$\psi_{xy}(t) = \psi_s(t-x)e^{-2\pi i y t}$$

so that $\psi_e = \psi_{x_0 y_0}$. Then form

$$I_u(x,y) = \left| \int_{-\infty}^{\infty} \psi_e(t)\overline{\psi}_{xy}(t)dt \right|^2$$

$$= |A_u(x_0-x, y_0-y)|^2$$

where $A_u(x,y)$ is called the ambiguity function of u and is explicitly given by

$$A_u(x,y) = \int_{-\infty}^{\infty} u(t - \tfrac{x}{2})\, \overline{u}(t + \tfrac{x}{2})\, e^{-2\pi i y t} dt.$$

The mapping A: u $\longrightarrow$ A_u is called the ambiguity operator. Then, as we will see in the next section,

$$I_u(x_0,y_0) = 1$$

$$I_u(x,y) \leqslant 1 \qquad \text{for all } x,y.$$

A device for finding the maximum of I_0, if it was unique, would determine x_0,y_0 and so r(0), v(0). Clearly the ideal shape for $A_u(x,y)$ would be the δ function at (0,0). But this can never be achieved.

2. ON THE RANGE OF THE AMBIGUITY OPERATOR

Because we can never achieve the perfect signal, we would like to be able to describe the range of the non-linear operator A in order to be able to search for signals with good properties. In this section we will see how basic facts about representation theory give a broad-brush picture of $A_u(x,y)$ and the range of A. We will then outline how nil-manifold theory can be used to obtain deeper results about the range of A.

It is elementary [A-T] to verify that if $f \in L^2(\mathbb{R})$ then $A_f(x,y) \in L^2(\mathbb{R}^2)$ and so we may view the ambiguity operator as a mapping

$$A: L^2(\mathbb{R}) \longrightarrow L^2(\mathbb{R}^2)\ .$$

To obtain a group theoretic setting for problem, we compose A with the unitary operator M on $L^2(\mathbb{R}^2)$ obtained by multiplying by $e^{\pi i x y}$. Explicitly, if

$$B = M \circ A\ .$$

Then

$$B_f(x,y) = \int_{-\infty}^{\infty} f(t)\bar{f}(t+x)e^{-2\pi iyt}dt .$$

It is clear that any information we obtain about B can be immediately translated to give information about A. Because of this, we will follow standard convention and also call B the ambiguity operator.

Now let N denote the 3-dimensional Heisenberg groups with multiplication given by

$$(x_1,x_2,x_3)(y_1,y_2,y_2) = (x_1+x_2y_1+y,x_3+y_3+x_1y_2) .$$

Let χ be the character on the maximal Abelian subgroup $\{(0,x_2,x_3)\in N\}$ given by

$$\chi(0,x_2,x_3) = e^{2\pi ix_3}$$

and let D be the irreducible representation of N on $L^2(\mathbb{R})$ obtained by inducing χ to N. If $f \in L^2(\mathbb{R})$ and < , > denotes the dot product in $L^2(\mathbb{R})$, we have

$$B_f(x,y) = \langle f,D(x,y,0)f\rangle .$$

By the standard results on positive definite functions on groups, we have the results at the end of Section 1; i.e.

1. $B_f(0,0) \geqslant 0$
2. $|B_f(x,y)| \leqslant B_f(0,0)$.

The two results below again follow from standard representation theory.

<u>**Theorem 1.**</u> If $f,g \in L^2(\mathbb{R})$ are such that $B(f) = B(g)$ then $f = \lambda g$ where λ is a constant of absolute value 1.

Theorem 2. Given $f,g \in L^2(\mathbb{R})$, there exists $h \in L^2(\mathbb{R})$ such that $B(f) + B(g) = B(h)$ if and only if $f = \lambda g$, λ a constant.

We now come to some results that use the explicit nature of N and not merely the fact that D is irreducible.

Let

$$\Gamma = \{(n_1,n_2,n_3)\in N \mid n_1,n_2,n_3 \in \mathbb{Z}\} .$$

Then on the compact nilmanifold $\Gamma\backslash N$ we have the regular representation R of N. Further, if $H \subset L^2(\Gamma\backslash N)$ is the character subspace with character $e^{2\pi i x_3}$ for the center of N, then $R|H$ is unitary equivalent to D and W defined by

$$W(f)(x_1,x_2,x_3) = e^{2\pi i x_3} \sum_{m\in\mathbb{Z}} f(x_1+m)e^{2\pi i m x_2}$$

is an intertwining operator.

Now for $f \in L^2(\mathbb{R})$ we say that f generates a L^2 basis of $L^2(\mathbb{R})$ provided

$$\mathcal{F} = \{f_{ab} = e^{2\pi i b t} f(t+a) \mid a,b\in\mathbb{Z}\}$$

has the following two properties:

1. $L^2(\mathbb{R})$ is the closure of finite linear combination of elements of $\mathcal{F}$.
2. No proper subset of $\mathcal{F}$ has this property.

Now consider $a \in \Gamma$ and $F \in H$. The fundamental theorem below follows from the fact that we can show that $e^{2\pi i(a_2y_1-a_1y_2)}W(f) = W(g)$ where $g(y_1,y_2,y_3) = D(a_1,a_2,0)f(y_1,y_2,y_3)$.

Theorem 3. Let $f \in L^2(\mathbb{R})$ generate an L^2 basis and let $\alpha(a,b)$ be a complex valued function on $\mathbb{Z}\times\mathbb{Z}$ with finite support. The range of the ambiguity operator A is the closure in $L^2(\mathbb{R}^2)$ of the set of functions

$$\sum_{a,b,c,d\in\mathbb{Z}} \alpha(a,b)\bar{\alpha}(c,d)K(a,b,c,d)A(f)(x+c-a,y+d-b)$$

where

$$K(a,b,c,d) = (-1)^{(a+c)(b+d)}e^{-\pi i((b+d)x-(a+c)y)} .$$

A proof can be found in [A-T].

3. SAMPLING THE AMBIGUITY FUNCTION.

Since computations are actually carried out digitally, it follows that we must approximate our functions by sampling them at a finite set of points and carrying out some computation on these finite sets of values. There will, of course, be an error incurred by this process and there are various choices of what to actually compute with our finite set of data so as to best approximate the continuous, or analog, model of our phenomena. We will now give a brief introduction to one method of sampling and computing an approximation to the ambiguity function.

In order to understand our approach to sampling ambiguity functions we will have to discuss a little more group theory. We begin by constructing the Heisenberg group for the group $\mathbb{Z}/N\mathbb{Z} = A$.

We define

$$N_A = \{(a_1,a_2,a_3) \mid a_i \in \mathbb{Z}/N\mathbb{Z}\}$$

and define multiplication by

$$(a_1,a_2,a_3)(b_1,b_2,b_3) = (a_1+b_1,a_2+b_2,a_3+b_3+a_1b_2).$$

It is clear that if A is any Abelian group and $\hat{A}$ is its dual group there is a natural Heisenberg group

$$N_A = \{(a,\hat{a},c) \mid a \in A,\ \hat{a} \in \hat{A},\ c \in \text{circle group}\},$$

but we will not need this general construction in this paper.

<u>Case</u> 1. Let $A = \mathbb{R}$, $\Gamma = \{(n_1,n_2,n_3) \in N_{\mathbb{R}} \mid n_\alpha \in \mathbb{Z}\}$, $\chi(0,x_2,x_3) = e^{2\pi i x_3}$ and $D = \mathrm{Ind}(\chi,B,N_{\mathbb{R}})$ where $B = \{(0,x_2,x_3)\} \subset N_{\mathbb{R}}$. D is well known to be irreducible.

<u>Case</u> 2. Let $A = \mathbb{Z}/M^2\mathbb{Z}$, $\Gamma = \{(n_1M,n_2M,0) \in N_A \mid n_1,n_2 \in \mathbb{Z}\}$, $\chi(0,a_2,a_3) = e^{2\pi i a_3/M^2}$ and $D_{M^2} = \mathrm{Ind}(\chi,B,N_A)$ where $B = \{(0,a_2,a_3)\} \subset N_A$. It is easily verified that D_{M^2} is irreducible.

Now consider the regular representation R_A of N_A on $L^2(\Gamma\backslash N_A)$ and let $H_\chi \subset L^2(\Gamma\backslash N_A)$ be defined by $f(a_1,a_2,a_3) \in H_\chi$ if and only if

$$f(a_1,a_2,a_3+c) = \chi(0,0,c)f(a_1,a_2,a_3).$$

Then $R_A \mid H_\chi = R_A^{\#}$ is irreducible and

$$R_{\mathbb{R}}^{\#} \approx D$$

$$R_{\mathbb{Z}/M^2\mathbb{Z}}^{\#} \approx D_{M^2}$$

where $\approx$ denotes unitary equivalence. Explicit intertwining operators can be given as follows. For $f \in L^2(\mathbb{R})$ define

$$W(f)(x_1,x_2,x_3) = e^{2\pi i x_3} \sum_{n=-\infty}^{\infty} f(x_1+n)e^{2\pi i n x_2}.$$

Then W intertwines $R_{\mathbb{R}}^{\#}$ and D. Similarly for $g \in L^2(\mathbb{Z}/M^2\mathbb{Z})$

$$W_{M^2}(g)(a_1,a_2,a_3) = e^{2\pi i a_3/M^2} \sum_{n=0}^{M-1} g(nM+a_1)e^{2\pi i a_2 N/M}$$

intertwines $R^{\#}_{\mathbb{Z}/M^2\mathbb{Z}}$ and D_{M^2}. Notice

$$B_{M^2}(g)(a_1,a_2) = \langle g, D_{M^2}(a_1,a_2,0)g\rangle.$$

We now introduce two subgroups of $N_{\mathbb{R}}$.

$$\Gamma(1/M) = \{(a/M, b/Mh, c/M^2) \in N_{\mathbb{R}} \mid a,b,c \in \mathbb{Z}\}$$

and

$$\Gamma(M) = \{(Ma, Mb, c) \in N_{\mathbb{R}} \mid a,b,c \in \mathbb{Z}\}.$$

One verifies that

$$\Gamma(M)\backslash\Gamma(1/M) \approx N_{\mathbb{Z}/M^2\mathbb{Z}}$$

In $\Gamma\backslash N_{\mathbb{R}}$ consider the subset

$$Y = \Gamma\backslash\Gamma(1/M).$$

Define

$V_s = \{F \mid Y \mid Y \in R^{\#}_{\mathbb{R}},\ F \text{ continuous}\}$
Then V_s is an M^2 dimensional vector space and $\Gamma(M)$ acting on V_s induces a representation of $\Gamma(M)\backslash\Gamma(1/M)$ on V_s that is $R^{\#}_{\mathbb{Z}/M^2\mathbb{Z}}$.

<u>**Theorem**</u> <u>**4.**</u> Let $g \in L^2(\mathbb{Z}/M^2\mathbb{Z})$ and $f \in L^2(\mathbb{R})$ be such that

$$W_{M^2}(g) = F \mid Y$$

and

$$W(f) = F.$$

Then

$$g(nM+a) = \sum_{k=-\infty}^{\infty} f(n + a/M + kM), \ 0 \leqslant a,n < M.$$

We may now form

$$B_{M^2}(g)(c,d) \text{ and } B(f)(c/M,d/M),$$

$c,d = 0,\ldots,M^2-1.$

It follows from a deep theorem of L. Richardson [R] that as N goes to infinity

$$B_{(MN)^2}(cN,dN) \longrightarrow B(f)(cN/MN,dN/MN)$$

$c,d = 0,\ldots,M^2-1.$

A proof of this will appear in a subsequent paper.

REFERENCES

[A-T] L. Auslander and R. Tolimieri, "Radar ambiguity functions and group theory," SIAM J. Math, Anal., 16(1985) 577-601.

[H-S] H.M. Helsaker and W. Schempp, "Radar detection, quantum mechanics and nilpotent harmonic analysis," Preprint.

[R] L. Richardson, "A class of idempotent measures," Acta. Math. 135 (1975) 121-154.

[W] E.P. Wigner, "On the quantum correction for thermodynamic equilibrium," Physics Review 40 (1932) 749-759.

[Wi] C.H. Wilcox, "The synthesis problem for radar ambiguity functions," MRC Technical Report 157 (1960), Mathematics Research Center, U.S. Army, University of Wisconsin.

[Wo] P.M. Woodward, "Probability and Information Theory, with applications to Radar," Pergamon Press, 1953.

KIRILLOV ORBITS AND DIRECT INTEGRAL DECOMPOSITIONS ON CERTAIN QUOTIENT SPACES

Lawrence Corwin*

Rutgers University
New Brunswick NJ 08903

Dedicated to George W. Mackey

§1.

A fundamental problem in representation theory is that of describing the unitary representations of a locally compact group G as a direct integral of simpler representations. The "abstract" problem (i.e., the problem of determining when such decompositions exist) takes up a good part of Mackey's Chicago lecture notes ↑16←; it is shown there that if G is Type I, then the problem has a satisfactory solution. The "concrete" problem (given G and a representation ρ, decompose ρ) can be considerably harder. For instance, if G is a semisimple Lie group and Γ is a discrete, cocompact subgroup, then the quasi-regular representation of G on $L^2(\Gamma\backslash G)$ is known to be a direct sum of irreducibles, each occurring with finite multiplicity, but little is known about which irreducibles appear. (See ↑6←.)

This paper will deal with what might be called a "very concrete version" of the problem: given certain G and ρ, show that ρ is equivalent to a direct integral of irreducibles of G <u>and</u> find an explicit intertwining operator giving the equivalence. To begin with, I shall describe the groups and the representations involved; then I shall describe the sort of explicit information desired; and finally, I shall try to explain one motivation for theorems of this sort.

---

* Supported by NSF Grant DMS 84-02704-01.

We look at connected, simply connected nilpotent Lie groups (hereafter, "nilpotent Lie groups"), G. Thanks primarily to Kirillov [13], we know a great deal about representation theory for these groups. Let K be a connected closed subgroup of G, and let ρ be the quasi-regular representation of G on $\mathcal{L}^2(K\backslash G)$; note that $K\backslash G$ has a G-invariant measure. In [13], Kirillov noted that ρ has a direct integral decomposition that can be described by his orbital picture, but he gave no details; recently, G. Grelaud [8], and independently Fred Greenleaf and I [5], gave an explicit direct integral decomposition of ρ (and, indeed, of $\mathrm{Ind}_{K\to G}\sigma$ for any $\sigma \in K^\wedge$). In general, ρ may have infinite multiplicity. Since it is easier to decompose representations of uniform multiplicity 1, we introduce a trick and an assumption. Let H be a (connected, closed) subgroup of G that normalizes K; we assume that $K \subsetneq H$. The H acts on $K\backslash G$ on the left, and we thus get a representation λ of H on $\mathcal{L}^2(K\backslash G)$:

$$(\lambda(h)f)\ (Kx) = f(Kh^{-1}x),\ h \in H \text{ and } x \in G.$$

Since λ and ρ commute, we have a representation $\lambda \times \rho$ of $H \times G$; of course, $\lambda | K$ is trivial. (In what follows, we identify representations of H/K with the corresponding representations of H.) We assume in what follows that $\lambda \times \rho$ is multiplicity-free and we say that (G,H,K) is a <u>multiplicity-free triple</u>. If, for example, K = {e}, then we can take H = G; as is well-known (see [22]), $\lambda \times \rho$ is then multiplicity-free for all general unimodular G. In fact, the explicit decomposition of $\lambda \times \rho$ amounts to the Plancherel theorem. (Similar remarks apply if K is normal.) For reference, we describe the situation of $G \times G$ acting on $\mathcal{L}^2(G)$:

$$\lambda \times \rho \overset{A}{\cong} \int_{G}^{\oplus} (\bar{\pi} \times \pi)\, d\mu(\pi),$$

where μ is Plancherel measure, and $\bar{\pi}$ is the contragredient of π.

More explicitly, we may regard $\bar{\pi} \times \pi$ as acting on $HS(\mathcal{H}_\pi)$, the (Hilbert) space of Hilbert-Schmidt operators on the Hilbert space where π operates, and then $f \in \mathcal{L}^1(G) \cap \mathcal{L}^2(G)$ is taken by A into the field $(\pi(f)) \in \int_{G}^{\oplus} HS(\mathcal{H}_\pi) d\mu(\pi)$. For sufficiently nice f, we also get Fourier inversion: $\pi(f)$ is of trace class a.e., and $f(e) = \int_G \mathrm{Tr}\pi(f) d\mu(\pi)$. When G is a nilpotent Lie group (with Lie algebra $\mathfrak{g}$) and $f \in \mathcal{S}(G)$, one can compute $\mathrm{Tr}\pi(f)$ by integrating $(f \circ \exp)$ over the Kirillov orbit in $\mathcal{O}_\pi$ in $\mathfrak{g}$ corresponding to $\pi \in G\hat{}$.

We shall try to mimic this situation for multiplicity-free triples in nilpotent Lie groups. This requires some notation. Let $\mathfrak{g}$, $\mathfrak{h}$, $\mathfrak{k}$ be the Lie algebras of G, H, K respectively. The elements of $G\hat{}$ (the unitary dual of G) are parametrized by the $Ad^*(G)$-orbits in $\mathfrak{g}^*$ by Kirillov theory. These orbits live in certain stratified sets, as follows: there are sets $U_1,\ldots,U_k$, $V_1,\ldots,V_k \subseteq \mathfrak{g}^*$, such that

(a) The U_j partition $\mathfrak{g}^*$;

(b) U_1 is Zariski-open in $\mathfrak{g}$, and U_j is Zariski-open in the variety $\mathfrak{g}^* \setminus (\bigcup_{i=1}^{j-1} U_i)$;

(c) V_j is the intersection of an algebraic variety with U_j, $\forall_j$;

(d) $Ad^*(G)V_j = U_j$;

(e) Distinct points in V_j belong to distinct $Ad^*(G)$-orbits;

(f) All orbits in U_j have the same dimension (d_j, say), and there is a rational map $F: \mathbb{R}^{d_j} \times V_j \longrightarrow \mathfrak{g}^*$ such that for $\ell \in V_j$, $F(\mathbb{R}^{d_j} \times \{\ell\}) = Ad^*(G)\ell$; for fixed ℓ, F is polynomial on $\mathbb{R}^d$.

Thus, each U_j is a union of the Ad^*-orbits parametrized by V_j. (The essence of the proof is in Chapter II, Section 1 of [18].)

Of course, similar remarks apply to $H^\wedge$, where there are sets U_j^o, V_j^o (say). Let $P:\mathfrak{g}^* \longrightarrow \mathfrak{h}^*$ be the canonical projection. We say that $\pi \in G^\wedge$ lies over $\sigma \in H^\wedge$ if $P(\mathcal{O}_\pi)$ meets $\mathcal{O}_\sigma$.

Denote the annihilator of $\mathfrak{k}$ in $\mathfrak{g}^*,\mathfrak{h}^*$ by $\mathfrak{k}^\perp(\mathfrak{g})$, $\mathfrak{k}^\perp(\mathfrak{h})$ respectively. There exists a unique integer j_0 such that $\mathfrak{k}^\perp(\mathfrak{g}) \cap U_{j_0}$ is Zariski-open and nonempty; similarly, there exists a unique j_1 such that $\mathfrak{k}^\perp(\mathfrak{h}) \subset U_{j_1}^0$ is Zariski-open and nonempty. We say that a property of elements of $G^\wedge$ is generic for (G, H, K) if there are Zariski-open subsets U, U^0 of $\mathfrak{k}^\perp(\mathfrak{g})$, $\mathfrak{k}^\perp(\mathfrak{h})$ respectively such that the property holds for $\pi \in G^\wedge$ whenever $\mathcal{O}_\pi$ meets $U \cap U_{j_0}$ and $P(\mathcal{O}_\pi)$ meets $U^0 \cap U_{j_i}^0$. (Similarly, we can speak of generic properties of $\ell \in \mathfrak{k}^\perp(\mathfrak{h})$, $\ell^0 \in \mathfrak{k}^\perp(\mathfrak{h})$, etc.) As an example, it follows from results of [8] or [5] that $\lambda \times \rho$ is multiplicity-free iff for generic $\ell \in \mathfrak{k}^\perp(\mathfrak{g})$, $Ad^*(G)\ell \cap \mathfrak{k}^\perp(\mathfrak{g}) = Ad^*(H)\ell$.

An orbital decomposition of $\lambda \times \rho$ (or of $(K\backslash G, H)$ will consist of the following data for generic $\sigma(K\backslash H)^\wedge$ and generic π lying over σ:

1. A complementary subspace $\mathfrak{s} = \mathfrak{s}(\sigma, \pi)$ to $\mathfrak{h}$ in $\mathfrak{g}$ and a map β_0: $\mathfrak{s} \longrightarrow G$ such that $S = \beta_0(\mathfrak{s})$ is a cross-section for $H\backslash G$; $\mathfrak{s}$, β_0 depend rationally on (σ, π); and β_0 carries Lebesgue measure to a G-invariant measure on $H\backslash G$.
2. A realization of π on a Hilbert space $\mathcal{H}_{\sigma,\pi}$ of functions from G to $\mathcal{H}_\sigma$, such that π acts by right translation (e.g., if $Ind_{H\to G}\sigma$ is irreducible, then the standard realization of the induced representation meets this description).
3. A Borel measure ν on the space E of such pairs (π, σ).

These satisfy certain conditions, to be listed below. Ideally, β_0 whould be exp, but this may not always be possible. (In fact, it does not seem to be known whether there is a cross-section $\mathfrak{s}$ to $\mathfrak{h}$ in $\mathfrak{g}$ such that $G = \exp \mathfrak{h} \exp \mathfrak{s}$.)

Note that $(\mathfrak{k}\backslash\mathfrak{h}) \times \mathfrak{s}$ is isomorphic to $\mathfrak{k}\backslash\mathfrak{g}$ in an obvious way. Let β: $(\mathfrak{k}\backslash\mathfrak{h}) \times \mathfrak{s} \longrightarrow K\backslash G$ be defined by $\beta(Y,W) = \exp(Y)\beta_0(W)$. We fix Lebesgue measure on $\mathfrak{k}\backslash\mathfrak{h}$ and lift it to get Haar measure on $K\backslash H$; we identify measures on S and $\mathfrak{s}$ via β_0, and we put Lebesgue measure on $\mathfrak{s}$ to a fixed G-invariant measure on $K\backslash G$. Of course, $\beta = \beta(\sigma, \pi)$ depends on σ and π.

We require the following of $\mathfrak{s}$, ν, and the realization $\mathcal{H}_{\sigma,\pi}$:

(a) $\lambda \times \rho \cong \int^{\oplus} \bar{\sigma} \times \pi \, d\nu(\sigma, \pi)$, the map corresponding to a decomposition A: $\int_E^{\oplus} HS(\mathcal{H}_\sigma, \mathcal{H}_{\sigma,\pi}) d\nu(\sigma, \pi) \cong \mathcal{L}^2(K\backslash G)$.

Here, $HS(\mathcal{H}_1,\mathcal{H}_2)$ is the Hilbert space of Hilbert-Schmidt operators from $\mathcal{H}_1$ to $\mathcal{H}_2$, and $\bar{\sigma}$ is the contragredient of σ.

(b) For $f \in \mathcal{S}(K\backslash G)$ (the Schwartz space of $K\backslash G$), $Af(\sigma, \pi)$ maps $\mathcal{H}_\sigma$ to continuous functions in $\mathcal{H}_{\sigma,\pi}$, so that $Af(\sigma, \pi)(x)$ is defined (by $Af(\sigma, \pi)(x)(v)) = (Af(\sigma, \pi)v)(x)$; $Af(\sigma, \pi)(x)$ is trace class for all x.

(c) For f as above.

$$f(e) = \int_E Tr(Af)(\sigma, \pi)(e) \, d\nu(\sigma, \pi),$$

and

$$\mathrm{Tr}(Af)(\sigma, \pi)(e) = \int_{\mathfrak{k}^{\perp}\cap\mathcal{O}_{\pi}\cap P^{-1}(\mathcal{O}_{\sigma})} (f\circ\mathcal{B})\hat{\ }(\ell)d\ell,$$

where $d\ell$ is an $\mathrm{Ad}^*(H)$-invariant measure on $\mathfrak{k}^{\perp}\cap\mathcal{O}_{\pi}\cap P^{-1}(\mathcal{O}_{\sigma})$ (this determines $d\ell$ up to a scalar multiple). We say that $\lambda \times \rho$ has a strong orbital decomposition if $\mathfrak{s}$ (hence $\mathcal{B}$) can be chosen to be independent of σ and π. In the future, we write $\mathcal{O}_{\pi,\sigma}$ for $\mathcal{O}_{\pi} \cap P^{-1}(\mathcal{O}_{\sigma})$ and $\mathcal{O}'_{\pi,\sigma}$ for $\mathcal{O}_{\pi,\sigma} \cap \mathfrak{k}^{\perp}$.

Note that if $K = \{e\}$ (or K is normal), then the Kirillov theory gives the orbital decomposition (with $H=G$). Thus, the orbital decomposition represents a generalization of part of Kirillov theory to homogenous spaces.

The main results of this paper involve proving the existence of orbital decompositions (or strong orbital decompositions) for $(K\backslash G, H)$ under various hypotheses on K, G, H. These theorems are given in Sections 3 and 4. Here, I shall try to explain the motivation behind these theorems (besides the obvious one, that Kirillov theory is pretty and ought to be exploited). It comes from partial differential equations.

Let $\mathfrak{u}(\mathfrak{g})$ denote the universal enveloping algebra of $\mathfrak{g}$. If G acts smoothly on a C^{∞} manifold M and preserves a measure ν on M, then the action gives a representation ρ of G on $L^2(M, \nu)$, and ρ extends to a representation of $\mathfrak{u}(\mathfrak{g})$ (also to be called ρ) on the space of C^{∞} vectors of ρ; this space includes $C_c^{\infty}(M)$. The operators in $\rho(\mathfrak{u}(\mathfrak{g}))$ are differential operators with C^{∞} coefficients, and we can use group representation theory to investigate their properties. Here, we consider local solvability. A differential operator L on M is said to be locally solvable at $m \in M$ if there exists an open neighborhood U of m such that for every $f \in C_c^{\infty}(U)$ we can find $u \in C^{\infty}(U)$ with $Lu = f$. We shall say that L is weakly locally solvable at m if for every $f \in C_c^{\infty}(U)$, there exists $u \in \mathcal{D}'(U)$ with $Lu = f$. As Hans Lewy showed in [14], not all differential operators are locally solvable; his example amounts to the case where G is the 3-dimensional Heisenberg group ($\mathfrak{g}$ = span (X, Y, Z), with [X, Y] = Z), M = G, ρ is the right regular representation, and L = X + iY. (In fact, this L is not weakly locally solvable; see [10].) Thus, sufficient conditions for local solvability of these operators are of some interest.

The basic procedure for proving local solvability via representation theory was sketched in [20] and developed in various other papers, such as [2] and [15]. (For surveys of results in the field, see [17] and [3].) The approach might be described as a "bootstrap" procedure. Suppose, for simplicity, that M = G and that ρ is the right regular representation. To show ρ(L) is locally solvable, one shows that σ(L) has a bounded left inverse A_σ for almost all $\sigma \in G^\wedge$ and that there are elements P, Q $\in \mathfrak{u}(\mathfrak{g})$ such that (a) $\|A_\sigma \, \sigma(P)\phi\| \; \mathcal{H} \; \|\sigma(Q)\phi\|$ for all C^∞ vectors ϕ (in $\mathcal{H}_\sigma$) and almost all σ, and (b) ρ(P) is locally solvable. Then it follows that ρ(L) is locally solvable. (While the versions of the theorem proved in the literature use the Plancherel theorem in the proof, the result actually depends only on direct integral theory.) Of course, one still has the problem of finding the locally solvable ρ(P).

The standard starting point is the following: suppose that M = K\G (K a connected, closed subgroup of G), and let D(K\G) be the algebra of differential operators on K\G commuting with the "right regular" representation ρ. If K = {e}, then D(G) = $\mathfrak{u}(\mathfrak{g})$, acting on the left. (For general K, the situation is discussed in [9] and [12]). Let ZD(K\G) be the center of D(K\G). When H = {e}, so that ZD(G) = $z\mathfrak{u}(\mathfrak{g})$m, then ZD(G) ≅ $\rho(z\mathfrak{u}(\mathfrak{g}))$ consists entirely of locally solvable operators (except for 0); this was proved in [19] by Raïs for nilpotent G, and by Duflo in [7] for the general case. (A different proof for solvable G was given by Rouvière, in [21].) In general, $\rho(z\mathfrak{u}(\mathfrak{g}))$ ZD(K\G), but there will be other elements as well. Is every (nonzero) element of ZD(K\G) locally solvable? In general, no, as Ron Lipsman has pointed out to me; his example is given in §6. But for nilpotent G the question is open, and it would also be interesting to know conditions on K which determine whether or not ZD(K\G) consists only of locally solvable elements. One relevant condition seems to be whether there exists H normalizing K such that the representation $\lambda \times \rho$ (discussed earlier) is multiplicity-free.

We now come back to orbital decompositions. Raïs's proof requires the Plancherel theorem and the Kirillov character formulas. If $(K\backslash G, H)$ has a strong orbital decomposition, then one can give a similar proof to show that every nonzero $L \in ZD(K\backslash G)$ is locally solvable. (Similar results appear to hold for orbital decompositions; I hope to treat this more general case in a future paper.) Thus, orbital decompositions provide the machinery for proving the local solvability of certain differential operators. In fact, they let one prove somewhat more: that these operators have tempered fundamental solutions.

Sections 3 and 4 of this paper are devoted to proving the existence of orbital decompositions in various cases. The main result of Section 3 is:

Theorem 1.1 Let (G, H, K) be a multiplicity-free triple of nilpotent Lie groups such that $K\backslash H$ is Abelian. Then $(K\backslash G, H)$ has an orbital decomposition.

As noted above, the hypothesis that $\lambda \times \rho$ is multiplicity-free has a geometric interpretation which enables one to decide its truth in specific situations; see [8] or [5]. The proof of Theorem 1.1 that is given in Section 3 actually gives a proof of the existence of orbital decompositions in a somewhat more general situation; the exact result is complicated, and is described at the end of Section 3. (I do not know of a multiplicity-free triple for which the proof of Theorem 1.1 does not apply.) The proof also gives the following "concrete" decomposition of $\lambda \times \rho$ whenever $\lambda \times \rho$ is multiplicity-free:

Theorem 1.2 Let (G, H, K) be a multiplicity-free triple; let E $= \{(\sigma, \pi)\colon \sigma \in (K\backslash H)^\wedge$ and $\pi \in G^\wedge$ lies above $\sigma\}$. Then there exist

(1) A measure ν on E;

(2) For each $(\sigma, \pi) \in E$, a realization of π on a Hilbert space

σ,π of functions from G to $\mathcal{H}_\sigma$ so that π acts by right translation;

(3) For each $(\sigma, \pi) \in E$, a map $A(\sigma, \pi)$: $\mathcal{S}(K\backslash G) \longrightarrow HS(\mathcal{H}_\sigma, \mathcal{H}_{\sigma,\pi})$; such that:

(a) The map A: $f \longrightarrow (A(\sigma, \pi)f) = (Af(\sigma, \pi))$ of $\mathcal{S}(K\backslash G)$ to $\int_E^\oplus Hs(\mathcal{H}_\sigma, \mathcal{H}_{\sigma,\pi})d\nu(\sigma, \pi)$ intertwines $\lambda \times \rho$ with $\int_E^\oplus \bar{\sigma} \times \pi \; d\nu(\sigma, \pi)$;

(b) A extends to an isometry of $\mathcal{L}^2(K\backslash G)$ with $\int_E^\oplus HS(\mathcal{H}_\sigma, \mathcal{H}_{\sigma,\pi})$.

Where Theorem 1.1 applies, the map A is the same as the map of this theorem.

Strong orbital decompositions are, of course, preferable to orbital decompositions; in Section 4, it is shown how to modify the proof of Theorem 1.1 to produce strong orbital decompositions in certain cases. Section 5 contains a theorem on local solvability of operators in ZD(K\G) for the case when there exists H such that (K\G, H) has a strong orbital decomposition. (I strongly suspect that this result can be extended to the case where there is simply an orbital decomposition, but I have not yet proved it.) Section 6 contains various examples and open questions. Section 2 is devoted to various small technical results, which are needed primarily in Sections 3 and 4.

Many of the results in this paper are an outgrowth of those in [4]. The conversations I had with Fred Greenleaf about that paper helped me greatly in clarifying my ideas about harmonic analysis on homogeneous spaces, and I am happy to acknowledge his influence on this paper.

§2.

The proof of Theorem 1.1 is by induction, and it turns out to be convenient to work in a rather general setting where we can consider a number of groups at the same time. This section is devoted to describing the setting and arranging notation.

In what follows, we shall often have to deal with maps $f: V \longrightarrow W$, where V and W are algebraic varieties and f is a rational function (strictly speaking, f is defined on a Zariski-open set of V; here and in what follows, Zariski-open sets will be assumed nonempty unless the contrary is stated). We shall say that a statement holds generically for f (or generically for V) if it holds on a Zariski-open subset of V; in particular, we shall usually assume that maps are defined only generically.

Let V be an algebraic variety; fix integers $K \leqslant d \leqslant n$. We consider a rational function $f: V \longrightarrow \mathbb{R}^{n^3}$ such that:

(1) For generic $\alpha \in V$, $f(\alpha)$ gives the structure constants for a nilpotent Lie algebra of dimension n (i.e., $f(\alpha) = (c^k_{i\,j}(\alpha))$, where, if $\{e_1, \ldots, e_n\}$ is the standard basis on $\mathbb{R}^n$, then

$$[e_i, e_j] = \sum_{k=1}^{n} c^k_{i\,j}(\alpha) e_k$$

makes R^n into a nilpotent Lie algebra);

(2) For generic $\alpha \in V$, the above Lie algebra structure makes $\mathbb{R}^d$ and $\mathbb{R}^k$ (regarded as embedded in $\mathbb{R}^n$ as the first d, k coordinates respectively) into subalgebras, with $\mathbb{R}^k$ and ideal of $\mathbb{R}^d$.

We shall write $\mathbb{R}^n$, $\mathbb{R}^d$, $\mathbb{R}^k$, with the Lie algebra structures given by $f(\alpha)$, as $\mathfrak{g}(\alpha)$, $\mathfrak{h}(\alpha)$, and $\mathfrak{k}(\alpha)$ respectively; the corresponding Lie groups are $G(\alpha)$, $H(\alpha)$, and $K(\alpha)$, identified as spaces with their algebras by exp. We call the $\mathfrak{g}(\alpha)$, $\mathfrak{h}(\alpha)$, $\mathfrak{k}(\alpha)$ <u>rationally varying families</u> of algebras. We shall say that f defines <u>rationally varying multiplicity-free</u> triples if $(G(\alpha), H(\alpha), K(\alpha))$ are generically multiplicity-free. We shall write [X, Y] for the bracket in all $\mathfrak{g}(\alpha)$ (using the context to specify the algebra under consideration).

If $A \in GL(\mathbb{R}^n)$, then A can be regarded as changing the basis of $\mathbb{R}^n$; A therefore induces a linear map on the structure constants (and thus on $\mathbb{R}^{n^3}$), which we call T_A. If A also maps $\mathbb{R}^d$ and $\mathbb{R}^k$ to themselves, then $T_A \circ f$ satisfies (1), (2) if f does, and $T_A \circ f$ defines algebraically varying multiplicity-free triples if f does. In what follows, we shall regard $f, T_{A(\alpha)} \circ f$ as equivalent.

We note the following simple facts about $\mathfrak{g}(\alpha)$, $\mathfrak{h}(\alpha)$, $\mathfrak{k}(\alpha)$:

1. The center of $\mathfrak{g}(\alpha)$ has constant dimension (generically). For $X = \sum_{i=1}^{n} a_i e_j$ is central if

$$0 = [x, e_j] = \sum_{i,k=1}^{n} (c^k_{i\,j}(\alpha)\, a_i) e_k$$

for all j, and the dimension of the common kernel of the maps $(a_i) \longmapsto \sum_{i=1}^{n} (c^k_{i\,j}(\alpha)) a_i$ for all j, k is generically constant. Similarly, the intersection of the center of $\mathfrak{g}(d)$ with $\mathfrak{h}(\alpha)$ or $\mathfrak{k}(\alpha)$ has constant generic dimension.

2. Similar remarks apply to other standard objects constructed from $\mathfrak{g}(\alpha)$–e.g., the normalizer of $\mathfrak{k}(\alpha)$, or $[\mathfrak{g}(\alpha), \mathfrak{g}(\alpha)]$.

3. Suppose that $Z(\alpha)$ is an element of $\mathfrak{k}(\alpha)$ that is central in $\mathfrak{g}(\alpha)$, let $z(\alpha)$ be the span of $Z(\alpha)$, and let $\bar{\mathfrak{g}}(\alpha)$ =

$\mathfrak{g}(\alpha)/z(\alpha)$, $\bar{\mathfrak{h}}(\alpha) = \mathfrak{h}(\alpha)/z(\alpha)$, $\bar{\mathfrak{k}}(\alpha) = \mathfrak{k}(\alpha)/z(\alpha)$. Then $\bar{\mathfrak{g}}(\alpha)$, $\bar{\mathfrak{h}}(\alpha)$, $\bar{\mathfrak{k}}(\alpha)$ are rationally varying families of algebras. For by a rational change of basis we may assume that $Z(\alpha) = e_1$; now if $\bar{f}: V \longrightarrow R^{(n-1)^3}$ is defined by $f(\alpha) = c_{i\,j}^{k}(\alpha)$: $2 \leqslant i, j, k \leqslant n$, then $\bar{f}$ defines the $\mathfrak{g}(\alpha)$, $\mathfrak{h}(\alpha)$, $\mathfrak{k}(\alpha)$. Similar remarks apply if $Z(\alpha)$ is central and is generically in $\mathfrak{h}(\alpha)\setminus\mathfrak{k}(\alpha)$ or generically in $\mathfrak{g}(\alpha)\setminus\mathfrak{h}(\alpha)$.

4. Suppose that $\mathfrak{g}(\alpha)$ has a 1-dimensional center $z(\alpha)$ generically; let $Y(\alpha)$ be in the second center of $\mathfrak{g}(\alpha)$ (so that $[\mathfrak{g}(\alpha), Y(\alpha)] \in z(\alpha)$), and assume that $Y(\alpha) \notin z(\alpha)$. Then it is a standard fact that $\mathfrak{g}_0(\alpha) = \{X \in \mathfrak{g}(\alpha)$: $[X, Y] = 0\}$ has codimension 1 generically. Let $\mathfrak{k}_0(\alpha) = \mathfrak{k}(\alpha) \cap \mathfrak{g}_0(\alpha)$, $\mathfrak{h}_0(\alpha) = \mathfrak{h}(\alpha) \cap \mathfrak{g}_0(\alpha)$. Then $\mathfrak{g}_0(\alpha)$, $\mathfrak{h}_0(\alpha)$, $\mathfrak{k}_0(\alpha)$ are rationally varying families of algebras. For if $\mathfrak{h}_0(\alpha) = \mathfrak{h}(\alpha)$ generally, then a rational change of basis lets us assume that $\mathfrak{g}_0(\alpha) = \text{span}\,\{e_j;\ 1 \leqslant j < n\}$, and the rest is simple. If $\mathfrak{h}_0(\alpha) \neq \mathfrak{h}(\alpha)$, but $\mathfrak{k}_0(\alpha) = \mathfrak{k}(\alpha)$ generically, then a rational change of basis lets us assume that $\mathfrak{g}_0(\alpha) = \text{span}\,\{e_j;\ j \neq d\}$. The remaining case is done in the same way.

In the course of the proof, we shall also need a simple calculation which we isolate here as a lemma.

Lemma 2.1 Suppose that G_0 has codimension 1 in G; let $X \in \mathfrak{g}\setminus\mathfrak{g}_0$ and write a typical element of G as $(s,y) = \exp sX \cdot y$, $y \in G_0$. Suppose further that π_0 is a unitary representation of G which is realized on a Hilbert space $\mathcal{H}_0$ of functions $f: G_0 \longrightarrow \mathbb{C}$, so that G acts by right translation. Let $y^s = \exp sX \cdot y \cdot \exp(-sX)$, $s \in \mathbb{R}$ and $y \in G_0$. Then one can realize $\pi = \text{Ind}_{G_0 \to G}\pi_0$ on the Hilbert space $\mathcal{H}$ of functions $f\colon G \longrightarrow \mathbb{C}$ such that:

(a) For every fixed s, the function $y \longrightarrow f(s,y^{-s}) = f_s(y)$ is in $\mathcal{H}_0$;

(b) $\int_{\mathbb{R}} \|f_s\|^2\, ds = \|f\|^2 < \infty$;

(c) The action of π is right translation.

Proof The standard realization of π is on the Hilbert space of functions $\phi\colon G \longrightarrow \mathcal{H}_0$ such that:

(a') $\phi(yx)(w) = \phi(x)(wy)$ $(= \pi_0(y)(\phi(x)(w)))$ for $y, w \in G_0$ and $x \in G$;

(b') $\int_{\mathbb{R}} \|\phi(\exp sX)\|^2 = \|\phi\|^2 < \infty$.

Define $f = f_\phi\colon G \longrightarrow \mathbb{C}$ by $f(x) = \phi(x)(e)$; it is easy to check that $\phi(x)(y) = f(yx)$. Thus, $\phi \longrightarrow f_\phi$ is injective. Moreover,

$$f(s,y^{-s}) = f(\exp sX \cdot y^{-s}) = f(y \exp sX) = \phi(\exp sX)(y);$$

hence $f_s \in \mathcal{H}_0$. Next, $\|\phi\|^2 = \int_{\mathbb{R}} \|f_s\|^2 ds$, as is easily checked. Let $\mathcal{H}$ be the space of f_ϕ, and transfer the action of π to $\mathcal{H}$. The lemma follows.

§3.

This section is devoted to the proof of the following generalization of Theorem 1.1:

Theorem 3.1 Suppose that we have rationally varying multiplicity-free triples of Lie algebras, $(\mathfrak{k}(\alpha), \mathfrak{h}(\alpha), \mathfrak{g}(\alpha))$, as described in Section 2, with $\mathfrak{k}(\alpha)\backslash\mathfrak{h}(\alpha)$ Abelian. Let $G(\alpha)$, $H(\alpha)$, $K(\alpha)$ be the corresponding groups. Then the $G(\alpha)$ have orbital decompositions in which the cross-sections $\mathcal{S}(\alpha, \sigma, \pi)$ vary rationally with α, σ, and π.

We shall consider a second theorem, produced from Theorem 1.2 as Theorem 3.1 is from Theorem 2.1 (but not explicitly stated here). We shall prove it along with Theorem 3.1.

The proof of Theorem 3.1 is rather long. To simplify the notation somewhat, we shall generally ignore the presence of α; the remarks in Section 2 suffice to show that our constructions can be made to depend rationally in α. As a result, property (b) of orbital decompositions will follow almost automatically from our construction, and we shall not verify it. In property (a), we shall verify the intertwining property of A only when the calculation presents some difficulty. We shall verify that A is an isometry into, but not that it is an isometry onto; this can be checked by approximation arguments like those in [4], or verified from more general considerations. In addition, we shall regularly ignore exceptional sets of elements (α,σ,π) which are proper Zariski-closed subspaces.

We make use of a preliminary decomposition. Let $\sigma \in (K\backslash H)^{\wedge}$, and let $\tau = \mathrm{Ind}_{H\to G}\sigma$; define A_σ: $\mathcal{L}^2(K\backslash G) \longrightarrow HS(H_\sigma, H_\tau)$ by

$$((A_\sigma f)(v))(x) = \int_{K\backslash H} f(hx)\sigma(h)^{-1}v\,dh,$$

$$v \in H_\sigma \text{ and } f \in \mathcal{L}^2(K\backslash G).$$

(Note: $\mathcal{H}_\tau$ is the space of functions ϕ: $G \longrightarrow \mathcal{H}_\sigma$ satisfying $\phi(hx) = \sigma(h)(\phi(x))$ and an L^2 condition.) It is shown in [4] that

$$\lambda \times \rho \cong \int_{(K\backslash H)^\wedge} \sigma \times \tau \, d\mu(\sigma)$$

(μ = Plancherel measure on $K\backslash H$),

under the map taking f to $(A_\sigma f)$. Note further that if we define $(A_\sigma f)(e)(v) = ((A_\sigma f)v)(e)$, the $(A_\sigma f)(e)$: $\mathcal{H}_\sigma \longrightarrow \mathcal{H}_\sigma$. We have the following trivial result.

Lemma 3.1 If $f \in \mathcal{S}(K\backslash G)$, then $A_\sigma f$ is a trace class operator, and

$$f(e) = \int_{(K\backslash H)} \operatorname{Tr}(A_\sigma f(e) \, d\mu(\sigma)$$

(μ = Plancherel measure on $(K\backslash H)$).

Proof $(A_\sigma f)(e) = \sigma(f^\vee)$, where $f^\vee(h) = f(h^{-1})$, and f is here restricted to h. Now the result is simply the Fourier inversion theorem for $K\backslash H$.

In proving Theorem 3.1, we shall generally decompose each $(A_\sigma f)$; that is, we shall try to write

$$\sigma \times \tau = \int_{V_\sigma} \sigma \times \pi \, d\mu_\sigma(\pi)$$

,

under a map realizing $A_\sigma f \longleftrightarrow (Af)(\sigma, \pi)$ so that

$$\operatorname{Tr}(A_\sigma f)(e) = \int_{V_\sigma} \operatorname{Tr} Af(\sigma, \pi) d\mu_\sigma(\pi)$$

Our inductive hypothesis will be that such decompositions exist for "smaller" rationally varying multiplicity-free triples.

There is one other lemma which we shall use in the proof of Theorem 3.1.

Lemma 3.2 Suppose that we have an orbital decomposition for $A_\sigma f$ in $(K\backslash H,G)$, with a cross-section for π given by $\beta = \beta_\pi$. Let $\gamma\colon x \longmapsto x^\gamma$ be an automorphism of G taking K to K and H to H, and let $\sigma^\gamma(x) = \sigma(x^\gamma)$, $\pi^\gamma(x) = \pi(x^\gamma)$. Then there is an orbital decomposition for $A_{\sigma^\gamma}f$ with the cross-section for π^γ given by β_π. Moreover, we have

$$A_{\sigma^\gamma}f(x) = A_\sigma(Uf)(x), \qquad Af(\sigma^\gamma,\ \pi^\gamma) = A(Uf)(\sigma,\ \pi),$$

where $Uf(x^\gamma) = f(x)$.

Proof We define ρ^γ on $\mathscr{L}^2(K\backslash G)$ by

$$(\rho^\gamma(x)f)(y) = f(yx^\gamma).$$

Then $\rho^\gamma \cong \rho$ under the intertwining operator U defined above; specifically,

$$\rho^\gamma = U\rho\ U^{-1}.$$

This same map takes $\mathcal{H}(\tau^\gamma)$ to $\mathcal{H}(\tau)$, where $\mathrm{Ind}_{H\to G}\sigma = \tau$ and $\mathrm{Ind}_{H\to G}\sigma^\gamma = \tau^\gamma$. Similarly, if π is realized on $\mathcal{H}_{\sigma,\pi}$ and $\pi(\gamma)$ is defined on $\mathcal{H}_{\sigma,\pi}$ by $(\pi(\gamma)(x)\phi)(y) = \phi(yx^\gamma)$, then we can define $\mathcal{H}_{\sigma^\gamma,\pi^\gamma}$ as the set of functions ϕ such that $x \longrightarrow \phi(x^\gamma)$ is in

$\mathcal{H}_{\sigma,\pi}$; the right action of G on $\mathcal{H}_{\sigma^\gamma,\pi^\gamma}$ gives a realization of π^γ. In this way, we can use U to transfer the orbital decomposition of σ (or of σ^γ) into the $\pi(\gamma)$ over to an orbital decomposition of σ^γ into the π^γ. We have

$$A_{\sigma^\gamma} f = A_\sigma(Uf);$$

correspondingly,

$$Af(\sigma^\gamma, \pi^\gamma) = A(Uf)(\sigma, \pi).$$

Thus

$$\mathrm{Tr}(A_{\sigma^\gamma} f)(e) = \mathrm{Tr}\, A_\sigma(Uf)(e) = \int_{E_\sigma} \mathrm{Tr}(AUf)(\sigma, \pi)(e)d\nu(e),$$

and, with $\delta = \gamma^{-1}$,

$$\begin{aligned}
\mathrm{Tr}(A(Uf))(\sigma, \pi)(e) &= \int_{\mathcal{O}'_{\pi,\sigma}} (Uf)\circ \beta)\hat{}(\ell)d\ell \\
&= \int_{\mathcal{O}'_{\pi,\sigma}} (f\circ \beta\circ \delta)\hat{}(\ell)d\ell \\
&= \int_{\mathcal{O}'_{\pi,\sigma}} (f\circ \beta)\hat{}(\gamma^*\ell)d\ell \\
&= \int_{\mathcal{O}'_{\pi,\sigma^\gamma}} (f\circ \beta)\hat{}(\ell)d\ell,
\end{aligned}$$

as desired.

We now give the proof of Theorem 3.1. We proceed by induction on (1) dim H\G, (2) dim K\H, and (3) dim G. If dim H\G = 0, then K is normal in G, and the result is the Plancherel theorem for G. Thus we assume that dim H\G and dim H are ≥ 1.

We proceed by examining cases. Since most of the argument does not require that $K\backslash H$ be Abelian, we do not make that assumption until it is needed; in particular, for most of the proof we do not assume that σ is 1-dimensional. In this way, we can prove Theorem 1.2 simultaneously with Theorem 3.1.

Case 1: There is a 1-parameter central subgroup $K_0 \subseteq K$. Then $\sigma(K_0)$ is trivial. Let $\mathfrak{k}_0$ be the Lie algebra of K_0; set $\bar{\mathfrak{g}} = \mathfrak{k}_0\backslash\mathfrak{g}$, $\bar{\mathfrak{h}} = \mathfrak{k}_0\backslash\mathfrak{h}$, and $\bar{\mathfrak{k}} = \mathfrak{k}_0\backslash\mathfrak{k}$, and let $\bar{G}$, $\bar{H}$, $\bar{K}$ be the corresponding Lie groups. It is easy to see that the theorem for G, H, K reduces to the theorem for $\bar{G}$, $\bar{H}$, $\bar{K}$.

Case 2: There is a 1-parameter central subgroup $H_0 \subseteq H \cap \operatorname{Ker} \sigma$, with $H_0 \not\subseteq K$. Let $\mathfrak{h}_0$ be the Lie algebra of H_0, and let $\bar{\mathfrak{g}}$ be a cross-section for $\mathfrak{h}_0\backslash\mathfrak{g}$ such that $\bar{\mathfrak{h}} = \bar{\mathfrak{g}} \cap \mathfrak{h}$ is a cross-section for $\mathfrak{h}_0\backslash\mathfrak{h}$ containing $\mathfrak{k}$. Let $\bar{G} = H_0\backslash G$, $\bar{H} = H_0\backslash H$; we may regard $\bar{\mathfrak{g}}$, $\bar{\mathfrak{h}}$ as the Lie algebras of $\bar{G}$, $\bar{H}$ respectively, and we may regard K as a normal subgroup of $\bar{H}$. Let $\bar{\sigma}$ be the representation of $\bar{H}$ corresponding to σ on H; similarly, let $\bar{\pi}$ be the representation of $\bar{H}$ corresponding to σ on H; similarly, let $\bar{\pi}$ (in $\bar{G}^\wedge$, lying over σ) correspond to π lying over σ, and let $\bar{\tau} = \operatorname{Ind}_{\bar{H}\to\bar{G}} \bar{\sigma}$. Then we have

$$\tau = \int_{V_\sigma} \pi_\alpha d\alpha, \quad \bar{\tau} = \int_{V_\sigma} \bar{\pi}_\alpha d\alpha,$$

as the direct integral decompositions of τ, $\bar{\tau}$ respectively.

For $f \in \mathcal{S}(K\backslash G)$, define $\bar{f} \in \mathcal{S}(K\backslash\bar{G})$ by $\bar{f}(\bar{x}) =$

$\int_{H_0} f(hx)dh$. By the inductive hypothesis, we have (for appropriate cross-sections, given by $\bar{\beta}$ on $\bar{H}\backslash\bar{G}$)

$$\mathrm{Tr}(\bar{A}\bar{f})\ (\bar{\sigma},\bar{\pi})(e) = \int_{\mathcal{O}'_{\bar{\pi},\bar{\sigma}}} (\bar{f}\circ\bar{\beta})\hat{}(\ell)d\ell$$

(where $\bar{A}$ is the map of Theorem 1.1 for $\bar{G}$)

$$= \int_{\mathcal{O}'_{\pi,\sigma}} (f\circ\beta)\hat{}(\ell)d\ell.$$

Let $(Af)(\sigma,\ \pi) = \bar{A}\bar{f}(\bar{\sigma},\bar{\pi})$. Then

$$\text{(3.1)}\qquad \mathrm{Tr}(Af)(\sigma,\ \pi)(e) = \int_{\mathcal{O}'_{\pi,\sigma}} (f\circ\beta)\hat{}(\ell)d\ell.$$

Furthermore, we may assume inductively that

$$\mathrm{Tr}(\bar{A}_{\bar{\tau}}f)(e) = \int_{V_\sigma} \mathrm{Tr}(\bar{A}\bar{f})(\bar{\sigma},\ \bar{\pi}_\alpha)(e)d\alpha,$$

as noted before Case 1; then

$$\text{(3.2)}\qquad \mathrm{Tr}(A_\tau f)(e) = \int_{V_\sigma} \mathrm{Tr}(Af)(\sigma,\pi_\alpha)(e)d\alpha$$

and (3.1) and (3.2) prove part (c) of the theorem for Case 2. Part (a) follows from the theorem for $(\bar{K},\ \bar{H},\ \bar{G})$ in the same way.

Case 3: There is a central 1-parameter subgroup not meeting H. Let Z be the center of G, $z = \log Z$, $\mathfrak{h}_1 = \mathfrak{h}\oplus z$. For K,

H_1, and G, the theorem holds (since $\dim(H_1\backslash G) < \dim(H\backslash G)$). Choose $Z_0 \in z$ with $Z_0 \neq 0$.

Given $\sigma \in (K\backslash H)^{\wedge}$, let σ_r be the representation of $K\backslash H_1$ given by

$$\sigma_r(h \exp zZ_0) = \sigma(h)e^{2\pi irz}.$$

Then if $\rho_r = \mathrm{Ind}_{H_1\to G}\sigma$, we have

$$\rho \cong \int_{\mathbb{R}}^{\oplus} \rho_r dr;$$

indeed, the map giving this isomorphism takes $\phi \in \mathcal{H}_\rho$ (ϕ is a function on G, with values in $\mathcal{H}_\sigma$) to $\phi^{\sim}$, a function on $G \times \mathbb{R}$, where

$$\phi^{\sim}(x, r) = \int_{\mathbb{R}} \phi(x \exp zZ_0)e^{-2\pi irz}\, dz.$$

(For fixed r, $\phi^{\sim}(k, r)$ lies in the space for ρ_r.) Thus

$$\begin{aligned}(A_\sigma f(e) &= \int_{K\backslash H} \sigma(h^{-1})f(h)dh \\ &= \int_{K\backslash H} \sigma(h^{-1}) \int_{\mathbb{R}} (\int_{Z} e^{-2\pi izr}f(h \exp zZ_0)dz)dr\ dh\end{aligned}$$

(Fourier inversion)

$$= \int_{\mathbb{R}} (\int_{K\backslash H} \sigma(h^{-1})e^{-2\pi izr} \int_{Z} f(h \exp zZ_0)dz)dr\ dh$$

$$= \int_Z \int_{K\backslash H_1} \sigma_r(h_1)^{-1} f(h_1) dh_1 \, dr$$

$$= \int_Z A_{\sigma_r}(f)(e) d\lambda.$$

Hence $\mathrm{Tr}(A_\sigma f)(e) = \int \mathrm{Tr}\ A_{\sigma_r}(f)(e)dr$. For each r and for each π lying over σ_r, the inductive hypothesis gives us a formula for $\mathrm{Tr}\ Af(\sigma_r, \pi)(e)$ and also shows that if

$$E_r = \{\pi \in G^\wedge : \pi \text{ lies over } \sigma_r\}$$

then

$$\int_{E_r} \|Af(\sigma_r, \pi)\|^2_{H-S} d\nu_r(\pi) = \|Af(\sigma_r)\|^2_{H-S}.$$

To complete the proof of Case 3, we need to verify that

$$\int_{\mathbb{R}} \|Af(\sigma_r)\|^2_{H-S} dr = \|Af(\sigma)\|^2_{H-S},$$

and we need to produce the cross-section corresponding to (σ_r, π). The first of these is a straightforward calculation from the definition; it amounts to the Plancherel Theorem for $\mathbb{R}$. For the second, if s_r is a cross-section corresponding to (σ_r, π) for $\mathfrak{g}$, $\mathfrak{h}_1$, $\mathfrak{k}$), then $s_r + z$ is one corresponding to (σ, π) for $(\mathfrak{g}, \mathfrak{h}, \mathfrak{k})$; if β_0' is the cross-section map for $\mathfrak{h}_1\backslash\mathfrak{g}$, define β_0 by $\beta_0(Y+W) = \exp Y\ \beta_0(W)$ for $W \in \mathfrak{s}_r$ and $Y \in Z$. Note that because Y is central, $\beta_0 = \exp$ if $\beta_0' = \exp$. The remaining details are easy to check.

If none of cases 1-3 apply, then G has 1-dimensional center $Z \subseteq$

H, and $\sigma | Z$ is nontrivial. We assume henceforth that this is so. Let $z^{(2)}$ denote the second center of $\mathfrak{g}$, and let z be the Lie algebra of Z.

Case 4: There is a nonzero element $y \in \mathfrak{k} \cap z^{(2)}$. Let $\mathfrak{g}_0 = \{X \in \mathfrak{g}: [X, Y] = 0\}$. Then $\mathfrak{g}_0$ has codimension 1 in $\mathfrak{g}$. Since $\mathfrak{h}$ normalizes $\mathfrak{k}$ and $\mathfrak{k}$ does not contain the center of $\mathfrak{g}$, $\mathfrak{h} \subseteq \mathfrak{g}_0$.

We now look at G_0, H, K, using a subscript $_0$ to denote objects belonging to this triple. A generic representation $\pi_0 \in G_0\hat{}$ lying over a generic $\sigma \in (H\backslash K)\hat{}$ induces to an irreducible $\pi \in G\hat{}$ lying over σ, since $Y \in \operatorname{Ker} \pi_0$ but $Y \notin \operatorname{Ker} \pi$. Thus if σ induces to τ_0 on G_0 and

$$\tau_0 = \int_{V_\sigma}^{\oplus} (\pi_0)_\alpha \, d\alpha,$$

then

$$\tau = \int_{V_\sigma}^{\oplus} \pi_\alpha \, d_\alpha, \; \pi_\alpha = \operatorname{Ind}_{G_0 \to G}(\pi_0)_\alpha.$$

Given $\pi_0 = (\pi_0)_\alpha$, let s_0 be the cross-section to $\mathfrak{h}$ in $\mathfrak{g}_0$, and let $s = s_0 + \mathbb{R}X$, where X is any element in $\mathfrak{g}\backslash\mathfrak{g}_0$. Write a typical element of G as $(y, s) = y \exp sX$, $y \in G_0$; for $f \in \mathcal{S}(K\backslash G)$, let $f_s(y) = f(y,s)$. We may realize π on the space $\mathcal{H}_\pi$ of functions ϕ: $G \longrightarrow \mathcal{H}_\sigma$ such that for each s, $y \longrightarrow \phi(Y, s)$ is in $\mathcal{H}_{\pi_0}$. Then

$$\pi \cong \int_{\mathbb{R}}^{\oplus} \pi_0 ds,$$

in an obvious way; define

$$Af(\sigma, \pi) = \int_{\mathbb{R}}^{\oplus} A_0 f_s(\sigma, \pi_0) ds,$$

where A_0 gives the decomposition for G_0. We now have all the ingredients for the orbital decomposition of $\lambda \times \rho$. It is easy to check that

$$\|A_0 f_s(\sigma, \pi)\|_{H-S}^2 = \int_{R} \|Af(\sigma, \pi_0)\|_{H-S}^2 ds,$$

and from this (a) follows; if $\mathcal{B}_0'$ is the cross-section map on $\mathfrak{s}'$ for $H\backslash G_0$, define $\mathcal{B}_0$ on $\mathfrak{s} = \mathfrak{s}'+\mathbb{R}X$ by $\mathcal{B}_0(W+tX) = \mathcal{B}_0'(W) \exp tX$. Only part (c) now presents any trouble. Since $\mathcal{O}_{\pi,\sigma}$ is the pre-image of $\mathcal{O}'_{\pi_0,\sigma}$ under the obvious projection of $\mathfrak{g}^*$ onto $\mathfrak{g}_0^*$, we have

$$\int_{\mathcal{O}'_{\pi,\sigma}} (f\circ \mathcal{B})^\wedge(\ell)d\ell = \int_{\mathcal{O}'_{\pi,\sigma}} (f_0\circ \mathcal{B}_0)^\wedge(\ell)d\ell,\ f_0 = f\big|_{G_0},$$

by Fourier inversion. On the other hand, our construction of $\mathcal{H}_\pi$ makes it clear that $Af(\sigma, \pi)(e) = A_0 f_0(\sigma, \pi_0)(e)$, and (c) follows immediately.

Case 5: $\mathfrak{k} \cap \mathfrak{z}^{(2)} = \{0\}$, but there exists a noncentral $Y \in \mathfrak{h} \cap \mathfrak{z}^{(2)}$. Again, let $\mathfrak{g}_0 = \{X \in \mathfrak{g}: [X, Y] = 0\}$. There are two subcases.

Subcase 5(a): $\mathfrak{h} \subseteq \mathfrak{g}_0$. We may assume (adding, if necessary, an element of z to Y) that $\sigma(Y) = 0$. Let $\mathfrak{k}' = \mathfrak{k} + \mathbb{R}Y$, and define $\bar{f}$ on $K'\backslash G$ by $\bar{f}(x) = \int_{\mathbb{R}} f(x \exp yY)dy$; now this subcase reduces in a straightforward way to Case 4 for G, H, K', and $\bar{f}$. We omit details.

Subcase 5(b): $\mathfrak{h} \not\subseteq \mathfrak{g}_0$. Let $\mathfrak{h}_0 = \mathfrak{h} \cap \mathfrak{g}_0$. By induction, the theorem holds for G_0, H_0, K. Let $\sigma \in (K\backslash H_0)^\wedge$ induce to σ, and let $\pi_0 \in \hat{G_0}$ lie over σ_0 and induce to π. Choose $X \in \mathfrak{h} \setminus \mathfrak{h}_0$, and let

$$y^t = (\exp tX)y(\exp -tX), \qquad t \in \mathbb{R} \text{ and } y \in G_0;$$

define σ_t by $\sigma_t(h) = \sigma(h^t)$ $(h \in H_0)$, and π_t by $\pi_t(y) = \pi_0(y^t)$, $y \in G_0$. For future reference, let

$$f_{s;t}(x_0) = f(x_0^t \cdot \exp sX), \qquad x_0 \in G_0 \text{ and } f \in \mathcal{S}(K\backslash G);$$

write f_s for $f_{s;0}$. Lemma 3.2 implies that

$$A(\sigma_{s-r}, \pi_{s-r})(f_t) = A(\sigma_s, \pi_s)(f_{t;r}). \tag{3.3}$$

To define $Af(\sigma, \pi)$, we first need descriptions of $\mathcal{H}_\sigma$ and $\mathcal{H}_\pi$. We may regard $\mathcal{H}_\sigma$ as $L^2(\mathbb{R}, H_{\sigma_0})$, with

$$(\sigma(h_0 \exp tX)\phi)(s) = \sigma_0(h_0)(\phi(s+t)).$$

We realize π' ($\cong\pi$) on the space $\mathcal{H}_{\pi'}$ of functions $\Phi : G \longrightarrow \mathcal{H}_{\pi_0}$, with

$$\text{(i)} \qquad \Phi(x_0x)(y_0) = \pi_0(x_0)(\Phi(x)(y_0)) = \Phi(x)(y_0x_0)$$

(x_0, $y_0 \in G_0$; $x \in G$), so that Φ is determined by its restriction to $\exp \mathbb{R}X$; and

$$\text{(ii)} \qquad \int_{\mathbb{R}} \|\Phi(\exp tX)\|^2 < \infty;$$

π' acts by right translation. Unfortunately, $\mathcal{H}_{\pi'}$ is not a space of functions from G to $\mathcal{H}_\sigma$. Thus define Q: $\mathcal{H}_{\pi'} \longrightarrow$ (functions from G to $\mathcal{H}_\sigma$) by

$$(Q\Phi)(x_0 \exp tX)(s) = \Phi(x_0{}^s \exp(s+t))(e).$$

Since $\Phi(x)(y_0) = \Phi(y_0x)(e)$, it is easy to see that Q is injective; we let $\mathcal{H}_\pi$ = image Q, and make $\mathcal{H}_\pi$ into a Hilbert space by declaring Q to be an isometry. Note that

$$(Q\Phi)(x_0\exp tX)(s) = (Q\Phi)(x_0^{-t})(s+t),$$

so that $F \in \mathcal{H}_\pi$ is determined by its restriction to G_0. Let $\pi = Q\pi'Q^{-1}$; then

$$(\pi(x)F)(y) = F(yx),$$

and we have the desired realization of π.

Since σ_t is realized on $\mathcal{H}_{\sigma_0}$, we know that $Ag(\sigma_t,\pi_t)(y)$:

$\mathcal{H}_{\sigma_0} \longrightarrow \mathcal{H}_{\sigma_0}$ for all $g \in \mathcal{S}(K\backslash G_0)$ and all $y \in G_0$. We define

$$((Af(\sigma, \pi)\phi)(x_0))(t) = \int_{\mathbb{R}} A(\sigma_s, \pi_s)f_{t-s}(x_0)(\phi(s))ds.$$

This definition may seem artificial, but in fact it is exactly what the constructions in [4] give for this case.

We first look at property (a). Note that

$$\begin{aligned} \lambda_0(h_0^{-1})f_{s+t;t}(x_0) &= f(h_0 \exp tX_0 \cdot x_0 \cdot \exp sX_0) \\ &= (\lambda(h_0 \exp tX_0)f)_s(x_0) \end{aligned}$$

(where λ_0 is the representation of H_0 on $\mathcal{L}^2(K\backslash G_0)$). Thus

$$(Af(\sigma, \pi)\bar{\sigma}(h_0 \exp rX)^{-1}\phi)(s_0)(t)$$

$$= \int_{R} A(\sigma_s, \pi_s)f_{t-s}(x_0)(x)(\sigma(h_0 \exp rX)\phi)(s)(ds)$$

$$= \int A(\sigma_s, \pi_s)f_{t-s}(x_0)(\sigma_s(h_0))(\phi(r+s))ds$$

$$= \int A(\sigma_s, \pi_s)(\lambda_0(h_0)f)_{t-s}(x_0)(\phi(r+s)))ds$$

(from the intertwining property of $A(\sigma_s, \pi_s)$)

$$= \int A(\sigma_{s-r}, \pi_{s-r})(\lambda_0(h_0)f)_{t-s+r}(x_0)(\phi(s))ds$$

$$= \int A(\sigma_s, \pi_s)(\lambda_0(h_0)f)_{t-s+r;r}(x_0)(\phi(s))ds$$

(from (3.3))

$$= A(\lambda(h_0 \exp rX)^{-1}f)\phi(x_0)(t),$$

which proves that $A(\sigma, \rho)$ intertwines $\bar{\sigma}$ and λ. The proof that $A(\sigma, \pi)$ intertwines π and ρ is similar. We shall omit the proof that $f \longrightarrow (A(\sigma, \pi)f)$ is an isometry; the argument is similar to that in subcase 6(c), where details are given, but the present case is somewhat easier.

For part (c), we first verify that $A(\sigma, \pi)f(x)$ is trace class. If $g \in \mathscr{L}^1(K\backslash H)$ and $f' \in \mathscr{S}(K\backslash G)$, then

$$A(\sigma, \pi)(\lambda(g)f')(x) = A(\sigma, \pi)f(x) \cdot \sigma(g).$$

But as in [25], we can find $g \in \mathscr{L}^1(K\backslash H)$ such that $\sigma(g)$ is trace class and $f = \lambda(g)f'$, $f' \in \mathscr{S}(K\backslash G)$. Now it is clear that $(A(\sigma, \pi)f)(x)$ is trace class. We have (see p. 102 of [24])

$$\operatorname{Tr} Af(\sigma, \pi)(e) = \int_{\mathbb{R}} \operatorname{Tr} Af_0(\sigma_s, \pi_s)(e)ds.$$

Let β be a cross-section for $H_0\backslash G_0$ corresponding to (σ_0, π_0). By Lemma 3.2,

$$\operatorname{Tr} Af_0(\sigma_s, \pi_s)(e) = \int_{\mathrm{Ad}^*(\exp sX)\,\mathcal{O}'_{\pi_0, \sigma_0}} (f_0 \circ \beta)^\wedge(\ell_0)d\ell_0$$

$$= \int_{\mathcal{O}'_{\pi_s, \sigma_s}} (f_0 \circ \beta)^\wedge(\ell_0)d\ell_0.$$

Hence

$$\operatorname{Tr} Af(\sigma, \pi)(e) = \int_{\mathbb{R}} ds \int_{\mathcal{O}'_{\pi_s, \sigma_s}} (f_0 \circ \beta)^\wedge(\ell_0)d\ell_0$$

$$= \int_{P(\mathcal{O}'_{\pi,\sigma})} (f_0 \circ \mathcal{B})^\wedge(\ell_0)d\ell_0$$

where P: $\mathfrak{g} \longrightarrow \mathfrak{g}_0^*$ is the canonical projection. Let ℓ be any extension of ℓ_0 to $\mathfrak{g}$, and let ℓ' satisfy $\ell'|_{\mathfrak{g}_0} = 0$, $\ell'(X) = 1$. Then

$$(f_0 \circ \mathcal{B})^\wedge(\ell_0) = \int_{\mathbb{R}} (f \circ \mathcal{B})^\wedge(\ell + t\ell')dt,$$

by a partial Fourier inversion. Since $\mathcal{O}_\pi = P^{-1}P(\mathcal{O}_\pi)$, the formula

$$\text{Tr } Af(\sigma, \pi)(e) = \int_{\mathcal{O}'_{\pi,\sigma}} (f \circ \mathcal{B})^\wedge(\ell)d\ell$$

holds and (c) is verified.

Case 6: $\mathfrak{g}^{(2)} \cap \mathfrak{h} = \mathfrak{z}$. We now choose $Y \in \mathfrak{z}^{(2)} \setminus \mathfrak{z}$, and let $\mathfrak{g}_0$ be the centralizer of Y, $\mathfrak{h}_0 = \mathfrak{g}_0 \cap \mathfrak{h}$. We show first that $\mathfrak{h}_0 \neq \mathfrak{h}$. Otherwise, Ad*(exp tY) is trivial on $\mathcal{O}_\sigma$ and takes $\mathcal{O}$ to itself, but if $\ell \in \mathcal{O}'_{\pi,\sigma}$ and $t \neq 0$, then

$$\text{Ad}^*(\exp tY)\ell \not\subset \text{Ad}^*(H)\ell.$$

(This last statement is simply the claim that $\mathfrak{r}(\ell) \cap \mathfrak{h} = \mathfrak{r}(\ell) \cap (\mathfrak{h} + \mathbb{R}Y)$, where $\mathfrak{r}(\ell)$ is the radical of ℓ.) It follows from [5] or [8] that $\lambda \times \rho$ is not multiplicity-free, a contradiction.

Thus, $\mathfrak{h}_0$ has codimension 1 in $\mathfrak{h}$. Let $\mathfrak{k}_0 = \mathfrak{g}_0 \cap \mathfrak{k}$; we get three subcases.

Subcase 6(a): $\mathfrak{k}_0 \neq \mathfrak{k}$. Let $X \in \mathfrak{k}-\mathfrak{k}_0$, let $K_0 = \exp \mathfrak{k}_0$ and $H_0 = \exp \mathfrak{h}_0$, and consider (G_0, H_0, K_0). The representations of $K_0 \backslash H_0$ are naturally equivalent to those of $K \backslash H$, and we shall not distinguish between them. Let $\pi \in G^\wedge$ lie above σ, and let $\pi_0 \in \hat{G_0}$ induce to π; define π_t by $\pi_t(w) = \pi_0(\exp tX \cdot w \cdot \exp(-tX))$, $w \in G_0$. Then the π_t all lie above σ (in $\hat{G_0}$). Suppose that $\mathcal{H}_0$ is the Hilbert space on which π_0 is realized; by assumption, $\mathcal{H}_0$ is a space of functions from G_0 to $\mathcal{H}_0$. We may realize π on $\mathcal{H}^\sim$ as in Lemma 2.1: $\mathcal{H}^\sim$ is the space of functions ϕ: $G \longrightarrow \mathcal{H}_\sigma$, with $\phi(s, w) = \phi(\exp sX \cdot w) = \phi_s(w)$, $w \in G_0$ and $s \in \mathbb{R}$, such that

(1) $w \longrightarrow \phi_s(w^{-s})$ (where $w^t = \exp tX \cdot w \cdot \exp(-tX)) \in \mathcal{H}_0$, all s, and

(2) $\|\phi\|_2^2 = \int \|\phi_s\|^2 ds < \infty$;

π acts by right translation. (Note that for each s, G_0 acts on the functions ϕ_s by π_s.) This realization of π is, however, not the one that we use in the theorem, since it does not lead to an orbital decomposition. (This is the case, in essence, because ρ acts on functions constant under the left action of K, while the functions in $\mathcal{H}^\sim$ are not controlled in the X-direction.)

We may assume (by scaling X) that $\pi_0([X, Y]) = 2\pi i I$. Let $w = w_0 \exp rY$, $r \in \mathbb{R}$; then for $\phi \in \mathcal{H}^\sim$, let $\phi_s(w) = \Phi_s(w^{-s})$. We have

$$\phi(s, w) = \Phi_s(w^{-s}) = \Phi_s(w_0^{-s} \exp rY^{-s})$$

(where $Y^t = \mathrm{Ad}(\exp tX)Y$)

$$= \Phi_s(w_0^{-s} \exp rY \exp rs[X,Y]) = \Phi_s(w_0^{-s})e^{2\pi irs}$$

$$(3.4) \qquad = \Phi_s(w_0)e^{2\pi irs} = e^{2\pi irs}\phi(s,w_0).$$

Now set $\phi^\wedge(\lambda, w) = (\mathcal{F}\phi)(\lambda, w) = \int_R \phi(s, w)e^{-2\pi is\lambda}ds$ and define $\pi^\wedge(x) = \mathcal{F}\pi(x)\mathcal{F}^{-1}$, $\mathcal{H}^\wedge = \mathcal{F}\mathcal{H}^\sim$. Since

$$(\pi(\exp tX\cdot y)\phi)(s, w) = \phi(s + t, w^{-t}y),$$

we see easily that for $\Phi \in \mathcal{H}^\wedge$,

$$\pi^\wedge(\exp tX\cdot y)\Phi(\lambda, w) = \Phi(\lambda,w^{-t}y)e^{-2\pi i\lambda t}.$$

Furthermore, (3.4) shows that $\phi(\lambda,w_0\exp rY) = \phi(\lambda-r, w_0)$. Thus Φ is determined by $\Phi(0,\cdot)$, a function from G to $\mathcal{H}_\sigma$. Define $\mathcal{H}$ to be the space of functions of the form

$$F(\exp sX\cdot w) = \Phi(0,w), \ \phi \in \mathcal{H}^\wedge,$$

and transfer the Hilbert space structure to $\mathcal{H}$ by setting $\|F\| = \|\Phi\| = \|\phi\|$ ($\Phi = \mathcal{F}\phi$). If we transfer the representation to $\mathcal{H}$, we get $\pi \cong \pi'$, where

$$\pi'(\exp tX\cdot y)F(\exp sX\cdot w) = \Phi(0,w^{-t}y) = F(\exp sX\cdot w\cdot \exp tX\cdot y).$$

Thus, $\mathcal{H}$ is a Hilbert space of functions on G, and $\pi' \cong \pi$ acts by translation.

It should be noted that this procedure is uniform for all π lying over σ, in that once we have picked Y, then the choice of π_0 is determined uniquely by π, the scaling of X is independent of π, and the maps from $\mathcal{H}$ to $\mathcal{H}'$ have the same formal definition. Let ρ^0

$= \mathrm{Ind}_{H_0 \to G_0}\sigma_t$, $\rho = \mathrm{Ind}_{H \to G}\sigma$; let σ_t be the representation on $H_1 = \exp \mathbb{R}Y \cdot H_0$ which is σ on H_0 and $2\pi it$ on Y, and let $\rho_t^0 = \mathrm{Ind}_{H_1 \to G_0}\sigma_t$. We have

$$\rho^0 = \int_{\mathbb{R}}^{\oplus} \rho_t^0 \, dt,$$

from induction in stages through H_1; furthermore, if $\rho_0^0 \cong \int_{V_0}^{\oplus} (\pi_\alpha)_0 d\alpha$, then

$$\rho_t^0 = \int_{V_0}^{\oplus} (\pi_\alpha)_t d\alpha,$$

where $(\pi_\alpha)_t(y) = (\pi_\alpha)_0(\exp tX \cdot y \cdot \exp(-tX))$; this is so because $\rho_t^0(y) = \rho_0^0(\exp tX \cdot y \cdot \exp(-tX))$. Finally,

$$\rho = \int_{V_0}^{\oplus} \pi_\alpha d\alpha, \ \pi_\alpha = \mathrm{Ind}_{G_0 \to G}(\pi_0)_\alpha.$$

We may assume (by induction) that the theorem holds for $(K_0 \backslash G_0, H_0)$; more specifically, we may assume that it holds for each ρ_t^0. Thus we have maps $A_0(\sigma_t, \pi_t)$: $f_0 \longrightarrow A_0 f_0(\sigma_t, \pi_t)$ for $(K_0 \backslash G_0, H_0)$. We now define $Af(\sigma, \pi)$. We know that $HS(\mathcal{H}_\sigma, \mathcal{H}^\sim)$ consists of operators

$$\int_{\mathbb{R}}^{\oplus} A_0 f(\sigma_t, \pi_t)dt = A'f(\sigma, \pi),$$

say; note that $K_0 \backslash G_0 \cong K \backslash G$. We defined an isomorphism $Q: \mathcal{H}^{\sim} \longrightarrow \mathcal{H}$ above; let

$$Af(\sigma, \pi) = QA'f(\sigma, \pi).$$

Now suppose that $\mathcal{B}$ is the proper cross-section for π_0. By Lemma 3.2, $\mathcal{B}$ is also a cross-section for each π_t. We have, therefore,

$$\operatorname{Tr} A_0 f(\tau_t, \pi_t)(e) = \int_{\mathcal{O}'_{\pi_t, \sigma_t}} (f \circ \mathcal{B})^{\wedge}(\ell)d\ell.$$

Thus

$$\int_{\mathbb{R}} \operatorname{Tr} A_0 f(\sigma_t, \pi_t)(e)dt = \int_{\mathcal{O}'_{\pi_t, \sigma_t}} (f \circ \mathcal{B})^{\wedge}(\ell)d\ell,$$

since $\mathcal{O}'_{\pi, \sigma}$ is the union of the $d\mathcal{O}'_{\pi_t, \sigma_t}$ and the measure is the obvious product measure. But from the definition of $(Af)(\sigma, \pi)$,

$$\int_{\mathbb{R}} \operatorname{Tr} Af(\sigma_t, \pi_t)(e)dt = \operatorname{Tr} Af(\sigma, \pi)(e).$$

This shows that (c) holds. For (a), note that

$$\|Af(\sigma, \pi)\|_{H-S}^2 = \|A'f(\sigma, \pi)\|_{H-S}^2$$
$$= \int_{\mathbb{R}} \|A_0 f(\sigma_t, \pi_t)\|_{H-S}^2 dt$$

and

$$\|Af(\sigma)\|^2_{H-s} = \int_{\mathbb{R}} \|Af(\sigma_t)\|^2_{H-s} dt;$$

the isometry property in (a) follows, and the intertwining property is a routine calculation. This completes the proof for the subcase 6(a).

Subcase 6(b): $\mathfrak{k}_0 = \mathfrak{k}$, $\mathfrak{h}_0 \neq \mathfrak{h}$, and $\sigma | H_0$ not irreducible. Let $X \in \mathfrak{h} \setminus \mathfrak{h}_0$; then let σ_0 induce to σ and let π_0 (lying over σ_0) induce to π. We are realizing π_0 on a space $\mathcal{H}_0$ of functions ϕ: $G_0 \longrightarrow K_0$ (where K_0 is the Hilbert space for σ_0). Then we can realize σ (in the usual way, using exp RX as a cross-section to $G_0 \setminus G$) in the space $K = \mathcal{L}^2(\mathbb{R}, \mathcal{H}_0)$, and we can form a representation $\pi^\sim$ of G on $\mathcal{H}^\sim$, the space of functions ϕ: $G \longrightarrow \mathcal{H}_0$ such that

(1) For fixed t,$y \longrightarrow \phi(y \exp tX) = \phi_t(y) \in \mathcal{H}_0$;

(2) $\int \|\phi_t\|^2 dt < \infty$;

$\pi^\sim \cong \pi$. (See Lemma 2.1.) We realize π on the space $\mathcal{H}$ of functions Φ: $G \longrightarrow K$ such that for all s, t $\in \mathbb{R}$ and all y $\in$ G,

(1) $\Phi(y \exp tX)(s) = \Phi(y^s \exp(s + t)X)(0)$, $y^s = \exp sX \cdot y \cdot \exp(-sX)$, so that Φ is determined by $\Phi(x)(0)$;

(2) $x \longrightarrow B\Phi(x) = \Phi(x)(0)$, with $\phi \in \mathcal{H}^\sim$.

We let $\mathcal{H}$ have the norm that makes B an isometry, and regard π as acting on $\mathcal{H}$ by right translation.

Now that we have π and σ, we can finish Subcase 6b. The details follow those of Subcase 4b, with no major changes. Here is a sketch: σ is realized on $K \cong \int_{\mathbb{R}}^{\oplus} K_t dt$, where K_t is the Hilbert space for σ_t, and a straightforward calculation shows that

$$\mathrm{Tr}(Af)(\pi,\sigma)(e) = \int_{\mathbb{R}} \mathrm{Tr}(A_0 f_0)(\pi_t,\sigma_t)(e)dt, \quad f_0 = f\,|_{K\backslash G_0}.$$

We compute $\mathrm{Tr}(A_0 f_0)(\pi_t, \sigma_t)(e)$ from the inductive hypothesis and Lemma 3.2; if P: $\mathfrak{g}^* \longrightarrow \mathfrak{g}_0^*$ is the obvious projection, then

$$\mathrm{Tr}(Af)(\pi,\sigma)(e) = \int_{P(\mathcal{O}'_{\pi,\sigma})} (f_0 \circ \beta)\hat{\ }(\ell)d\ell,$$

and we finish the proof by a calculation of a partial Fourier transform, as in Case 4.

Subcase 6(c): $\mathfrak{k}_0 = \mathfrak{k}$, $\mathfrak{h}_0 \neq \mathfrak{h}$, and $\sigma | H$ is not irreducible. The proof in this case begins like the one for case 6a. We define $\mathcal{H}^\sim$ as in case 6a and note that (3.4) still holds. Now, however, define

$$\phi\hat{\ }(\lambda, w) = (\mathcal{F}\phi)(\lambda, w) = \int_{\mathbb{R}} \sigma(\exp sX)(\phi(s, w))e^{-2\pi i\lambda s}ds,$$

Let $\mathcal{H}\hat{\ } = \mathcal{F}\mathcal{H}^\sim$, and let $\pi\hat{\ }(x) = \mathcal{F}\pi(x)\mathcal{F}^{-1}$. A calculation like that in case 6a gives

$$\pi\hat{\ }(\exp tX \cdot y)\phi\hat{\ }(\lambda,w) = \sigma(\exp tX)\phi\hat{\ }(\lambda, w^{-t}y))e^{2\pi i\lambda t}.$$

As in case 6a, ϕ is determined by values on $\{0\} \times G_0$. Define $\mathcal{H}$ to be the space of functions of the form

$$F(\exp sX \cdot w) = \sigma(\exp sX)(\phi(0, w)), \qquad \phi \in \mathcal{H},$$

and transfer the Hilbert space structure to $\mathcal{H}$ by letting $\|F = \|\phi\| = \|\phi\|$ when $\phi = F\phi$. Transferring $\pi\hat{\ }$ to $\mathcal{H}$ gives us π', where

$$\pi'(\exp tX \cdot y)F(\exp sX \cdot w) = F(\exp sX \cdot w \exp tX \cdot y).$$

Then π' is the desired realization of π.

Given f on $K\backslash G$, we let f^s be the function on $K\backslash G_0$ defined by

$$f^s(y) = f(\exp sX \cdot y).$$

Then $A_0 f^s(\sigma_t, \pi_t)$ is defined, where $\sigma_0 = \sigma\big|_{H_0}$ and σ_t, π_t are defined as in subcase 6(a); here, $\sigma_t = \sigma_0$. (The subscript is in A_0 because A_0 is the decomposition map for $(K\backslash H_0, G_0)$.) Define

$$A'f(\sigma_t, \pi_t) = \int_{R} A_0 f^s(\sigma_t, \pi_t)\sigma(\exp(t-s)X)ds \in HS(\sigma_t, \pi_t).$$

We have

$$\mathcal{H}^{\sim} \cong \int_{\mathbb{R}}^{\oplus} \mathcal{H}(\sigma_t, \pi_t)dt,$$

as in subcase 6(a); let

$$(A^{\sim}f)(\sigma, \pi) = \int^{\oplus} A'f(\sigma_t, \pi_t)dt.$$

As in subcase 6(a), we define Af from $A^{\sim}f$ by transferring the action from $\mathcal{H}^{\sim}$ to $\mathcal{H}$. Some calculation shows that for $y \in G_0$,

$$(Af(\sigma, \pi)v)(y) = \int_{\mathbb{R}}\int_{\mathbb{R}} \sigma(\exp(-tX))A_0 f^s(\sigma_t, \pi_t)(y)\sigma(\exp(t-s)X)v)dt\ ds;$$

note that $Af(\sigma, \pi)v$ is determined by its restriction to G_0. We shall check the properties of $Af(\sigma, \pi)$ in some detail, since the calculations are fairly complicated; they can also serve as a model for the calculations omitted in other cases. First, we collect some useful formulas. Set

$$f^{s;t}(y) = f(\exp(s-t)X \cdot y \cdot \exp tX), \qquad y \in G_0 \text{ and } x,\ t \in \mathbb{R},$$

as in the proof of subcase 5. From Lemma 3.2, we have

$$A_0 f^s(\sigma_{t+t'}, \pi_{t+t'})(y) = A_0 f^{s;t}(\sigma_{t'}, \pi_{t'})(y^t). \tag{3.5}$$

Next, a straightforward calculation shows the following: for $\phi \in \mathcal{H}_{\sigma,\pi}$, let $\phi^t(y) = \phi(\exp tX \cdot y)$, $y \in G_0$. Then $\phi^t \in \mathcal{H}_{\sigma_0, \pi_0}$, and one gets

$$Af(\sigma, \pi)^* \phi = \iint_{\mathbb{R}\mathbb{R}} \sigma(\exp(s-t)X) A_0 f^s(\sigma_t, \pi_t)^*(\phi^t)\, ds\, dt. \tag{3.6}$$

We now verify property (a). We have, for $y \in G_0$,

$$(\pi(\exp tX)Af(\sigma, \pi)v)(y) = Af(\sigma, \pi)v(\exp tX \cdot y^{-t})$$

$$= \sigma\ (\exp tX)(Af(\sigma, \pi)v(y^{-t})$$

$$= \iint_{\mathbb{R}\mathbb{R}} \sigma(\exp(t-t')X) A_0 f^s(\sigma_{t'}, \pi_{t'})(y^{-t}) \; \sigma(\exp(t'-s)X)v\ dt'\ ds$$

$$= \iint_{\mathbb{R}\mathbb{R}} \sigma(\exp -t'X) A_0 f^s(\sigma_{t'+t}, \pi_{t'+t})(y^{-t}) \; \sigma(\exp(t'+t-s)X)v\ dt'\ ds$$

$$= \iint_{\mathbb{R}\mathbb{R}} \sigma(\exp -t'X)A_0 f^{s;t}(\sigma_{t'}, \pi_{t'})(y)$$

$$\sigma(\exp(t'+t-s)X)v \; dt' \; ds$$

(by Lemma 3.2)

$$= \iint_{\mathbb{R}\mathbb{R}} \sigma(\exp -t'X)A_0 f^{s+t;s}(\sigma_{t'}, \pi_{t'})(y)$$

$$\sigma(\exp(t'-s)X)v \; dt' \; ds.$$

This last expression is

$$A(\rho(\exp tX)f)(\sigma, \pi)v(y),$$

which shows that $A(\sigma, \pi)$ intertwines $\rho(\exp tX)$ with $\pi(\exp tX)$. The proof that $A(\sigma, \pi)$ intertwines $\rho(x)$ with $\pi(x)$, $x \in G_0$, is similar but easier.

In showing the intertwining property of λ with $\bar{\sigma}$, the case of elements in H_0 is the more complicated one. We have, for $h \in H_0$,

$$(Af)(\sigma, \pi)\bar{\sigma}(h)^{-1}v(y)$$

$$= \iint_{\mathbb{R}\mathbb{R}} \sigma(\exp -tX)A_0 f^{s}(\sigma_{t}, \pi_{t})(y)$$

$$\sigma(\exp(t-s)X)\sigma(h)v \; dt \; ds$$

$$= \iint_{\mathbb{R}\mathbb{R}} \sigma(\exp -tX)A_0 f^{s}(\sigma_{t}, \pi_{t})(y)\sigma(h^{t-s})$$

$$\sigma(\exp(t-s)X)v \; dt \; ds$$

$$= \iint_{\mathbb{R}\mathbb{R}} \sigma(\exp -tX)A_0 f^{s}(\sigma_{t}, \pi_{t})(y)\sigma_{t}(h^{-s})$$

$$\sigma(\exp(t-s)X)v \; dt \; ds$$

$$= \iint_{\mathbb{R}\mathbb{R}} \sigma(\exp -tX)A_0(\lambda(h^{-s})^{-1}f^s)(\sigma_t, \pi_t)(y)$$

$$\sigma(\exp(t-s)X)v \ dt \ ds,$$

by the intertwining property of A_0. But

$$\lambda(h^{-s})^{-1}f^s(x) = f^s(h^{-s}x)$$

$$= f(h \exp sX \cdot x) = (\lambda(h)^{-1}f)^s(x),$$

so that this last expression is

$$\iint_{\mathbb{R}\mathbb{R}} \sigma(\exp(-tX)A_0(\lambda(h)^{-1}f)^s(\sigma_t, \pi_t)(y)\sigma(\exp(t-s)X)v \ dt \ ds$$

$$= (A(\lambda(h)^{-1}f)(\sigma, \pi)v)(y),$$

as required. The proof for $h = \exp tX$ involves a simple change of variables in the integral.

To finish off the proof of (a), we need to compute $\|Af(\sigma, \pi)\|^2_{H-S}$. From (3.5), we have

$$(Af(\sigma, \pi)Af(\sigma, \pi)^*)\phi(y)$$

$$= \iiiint_{\mathbb{R}\mathbb{R}\mathbb{R}\mathbb{R}} \sigma(\mathrm{Exp}\ -t'X)A_0f^s(\sigma_t, \pi_t)\sigma(\exp(s'-t'-s+t)X)$$

$$A_0f^{s'}(\sigma_{t'}, \pi_{t'})^*\phi(y) \ ds \ dt \ dx' \ dt'$$

$$= \iiiint_{\mathbb{R}\mathbb{R}\mathbb{R}\mathbb{R}} \sigma(\exp -tX)A_0f^s(\sigma_t, \pi_t)\sigma(\exp(t-s-t'+s')X)$$

$$A_0f^{s'}(\sigma_{t'}, \pi_{t'})^*\sigma(\exp tX)\phi(y) \ ds \ dt \ ds' \ dt'$$

$$= \int_{\mathbb{R}}\int_{\mathbb{R}}\int_{\mathbb{R}}\int_{\mathbb{R}} \tilde{\sigma}(\exp -tX)A_0f^s(\sigma_t, \pi_t)\sigma(\exp(t-s-t'+s')X)$$

$$A_0f^{s'}(\sigma_{t'}, \pi_{t'})^*\tilde{\sigma}(\exp t'X)\phi)(y) \, ds \, dt \, ds' \, dt',$$

where $\tilde{\sigma}(\exp tX)\phi$ is the function defined by

$$(\tilde{\sigma}(\exp t'X)\phi)(y) = \phi(\exp t'X \cdot y).$$

From the results on pp. 102, 105 of [24],

$$\|Af(\sigma, \pi)\|^2_{H-S}$$

$$= \int_{\mathbb{R}}\int_{\mathbb{R}} \mathrm{Tr}(\tilde{\sigma}(\exp -tX)A_0f^s(\sigma_t, \pi_t)$$

$$A_0f^s(\sigma_t, \pi_t)^*\tilde{\sigma}(\exp tX))ds \, dt$$

$$= \int_{\mathbb{R}}\int_{\mathbb{R}} \mathrm{Tr}(A_0f^s(\sigma_t, \pi_t)A_0f^s(\sigma_t, \pi_t)^*)ds \, dt.$$

Now integrate over all π lying over σ; recalling that $\sigma_t \cong \sigma_0$, we get

$$\int_{E_\sigma} \| Af(\sigma, \pi)\|^2_{H-S} d\nu(E)$$

$$= \int_{E_\sigma}\int_{\mathbb{R}}\int_{\mathbb{R}} \|A_0f^s(\sigma_0, \pi_t)\|^2_{H-S} ds \, dt \, d\nu(E)$$

$$= \int_{E_{\sigma_0}}\int_{\mathbb{R}} \|A_0f^s(\sigma_0, \pi_0)\|^2_{H-S} ds \, d\nu_0$$

(E_{σ_0} is the space of $\pi_0 \in G^\wedge$ lying over σ_0)

$$= \int_{\mathbb{R}} \|A_{\sigma_0} f^s\|_2^2 ds$$

(by the inductive hypothesis)

$$= \|A_\sigma f\|_2^2,$$

as a straightforward computation (using [4]) shows. This completes the verification of (a) (and the proof of Theorem 1.2).

It is when we come to verify (c) that we finally use the hypothesis that σ is scalar (or, at least, that $\sigma(X)$ is scalar). Let $\sigma(\exp tX) = e^{2\pi i \gamma t}$. We have, from the formula for $Af(\sigma, \pi)(e)$,

$$\operatorname{Tr} Af(\sigma, \pi)(e) = \iint_{\mathbb{R}\mathbb{R}} \operatorname{Tr}(\sigma(\exp sX)A_0 f^s(\sigma_t, \pi_t)(e))ds\ dt$$

$$= \iint_{\mathbb{R}\mathbb{R}} e^{2\pi i \gamma s} \operatorname{Tr} A_0 f^s(\sigma_t, \pi_t)(e)ds\ dt.$$

Let σ be the cross-section map for π_0, σ_0 (hence for π_t, σ_t); let β be the map for π, σ, using this same cross-section of $\mathfrak{h}\backslash\mathfrak{g} \cong \mathfrak{h}_0\backslash\mathfrak{g}_0$. If $y = \beta_0(W_1, W_2) \in K\backslash G_0$, then

$$\exp sX \cdot y = \beta(W_1 + sX + W_1'), \quad W' \in \mathfrak{k}\backslash\mathfrak{h}_0,$$

and the Campbell-Baker-Hausdorff formula shows that W' is in the ideal generated by $[X, \mathfrak{h}_0]$. Hence $\sigma_t(W') = 0$ for all t. In particular, $\ell_0(W') = 0$ if $\ell \in \mathfrak{h}_0^*$ is in the orbit corresponding to σ_t.

Define

$$f^\wedge(s, \ell_0) = \int f \circ \beta(sX + W_1, W_2) e^{2\pi i \ell_0(W_1 + W_2)} dW_1\, dW_2$$

for $s \in \mathbb{R}$ and $\ell_0 \in (\mathfrak{k}\setminus\mathfrak{g}_0)^*$. By the inductive hypothesis and the above remark about W',

$$\operatorname{Tr} A_0 f^s(\sigma_t, \pi_t)(e) = \int_{\mathcal{O}'_{\pi_t, \sigma_t}} f^\wedge(s, \ell_0) d\ell_0.$$

Let ℓ' be 1 on X and 0 on $\mathfrak{g}_0$; extend $\ell_0 \in (\mathfrak{k}\setminus\mathfrak{g}_0)^*$ to be 0 on X. Then

$$\int_{\mathbb{R}} e^{-2\pi i \gamma s} \operatorname{Tr} A_0 f^s(\sigma_t, \pi_t)(e) ds$$

$$= \int_{\mathbb{R}} \int_{\mathcal{O}'_{\pi_t, \sigma_t}} f^\wedge(s, \ell_0) e^{-2\pi i \gamma s} d\ell_0 \, ds$$

$$= \int_{\mathcal{O}'_{\pi_t, \sigma_t}} f^\frown(\ell_0 + \gamma\lambda') d\ell_0.$$

Now let $\mathcal{O}^\sim_{\pi, \sigma} = \bigcup_{t \in R} \mathcal{O}'_{\pi_t, \sigma_t} \in (\mathfrak{k}\setminus\mathfrak{g}_0)^*$.

Then $\mathcal{O}'_{\pi, \sigma} = \{\ell \in (\mathfrak{k}\setminus\mathfrak{g})^*: \ \ell(X) = \gamma, \ \ell\big|_{\mathfrak{g}_0} \in \mathcal{O}^\sim_{\pi, \sigma}$,

so that

$$\operatorname{Tr} Af(\sigma, \pi)(e) = \int_{\mathbb{R}} \int_{\mathcal{O}^\sim_{\pi_t, \sigma_t}} f^\frown(\ell_0 + \gamma\ell') d\ell_0 \, dt$$

$$= \int_{\mathcal{O}^\sim_{\pi_t, \sigma_t}} f^\frown(\ell_0 + \gamma\ell') d\ell_0$$

$$= \int_{\mathcal{O}'_{\pi, \sigma}} f^\frown(\ell) d\ell.$$

This proves case 6 and completes the proof of Theorem 3.1 (and hence of Theorem 1.1). As noted above, we have also proved Theorem 1.2.

A certain amount more can be obtained from this proof. Let $\sigma \in (K\backslash H)^\wedge$, and let π lie over σ. Say that $(K_1, H_1, G_1; \sigma_1, \pi_1)$, where $\sigma_1 \in (K_1\backslash H_1)^\wedge$ and π_1 lies over σ_1, is an <u>immediate</u> <u>reduction</u> of $(K, H, G; \sigma, \pi)$ if one of the following holds:

1. There is a 1-parameter central subgroup $Z \in \operatorname{Ker} \pi$, and

$$K_1 = PK, \quad H_1 = PH, \quad P \circ \pi_1, \quad \sigma = P \circ \pi_2,$$

where $P\colon G \longrightarrow G_1 = G/Z$ is the canonical projection;

or

2. There is a 1-parameter central subgroup $Z \in G$ with $Z \cap H = \{e\}$, and

$$K_1 = K, \quad H_1 = ZH, \quad G_1 = Z,$$
$$\pi_1 = \pi, \quad \sigma_1(h \exp tZ) = \sigma(h)\lambda(\exp tZ),$$

where $\pi(\exp tX) = \lambda(\exp tZ)I$;

or

G has 1-dimensional center $Z \subseteq H$, with $\sigma(Z) \lneqq I$, and one of the following holds:

3. There is a nonzero element $Y \in \mathfrak{h} \cap \mathfrak{g}^{(2)}$, and

$$K_1 = K \cap G, \quad H_1 = H \cap G_1,$$
$$\sigma \text{ lies over } \sigma_1, \quad \pi = \operatorname{Ind}_{G_1+G} \pi_1,$$

where G_1 is the centralizer of $\exp \mathbb{R}Y$;

or

4. There is a nonzero element $Y \in \mathfrak{g}^{(2)}\backslash \mathfrak{z}$, G_1 is the

centralizer of exp $\mathbb{R}Y$, and

a) $K_1 = K \cap G_1 \subsetneq K$, $H_1 = H \cap G_1$, σ lies over σ_1, $\pi = \mathrm{Ind}_{G_1 \to G}$

or

b) $K_1 = K$, $H_1 = H \cap G_1 \subsetneq H$, σ lies over σ_1, $\pi = \mathrm{Ind}_{G_1 \to G}\pi_1$, and either $\sigma = \mathrm{Ind}_{H_1 \to H}\sigma_1$

or

$\sigma|_{H_1} = \sigma_1$ and there exists $Y \in \mathfrak{h} \setminus \mathfrak{h}_1$ such that $\sigma(Y)$ is scalar.

Say that (K', H', G' σ', π') is a <u>reduction</u> of (K, H, G, σ, π) if it can be obtained from (K, H, G, σ, π) by a chain of immediate reductions.

Theorem 3.2 If for generic $\sigma \in (K\setminus H)^\wedge$ and generic $\pi \in G^\wedge$ lying over σ there exists a reduction (K', H', G', σ', π') with K' normal in G', then (K\H, G) has an orbital decomposition.

A reader with any remaining strength can verify that the proof of Theorem 3.1 also amounts to a proof of Theorem 3.2.

§4.

In a number of interesting cases, the cross-sections s (and the map β) of Theorem 1.1 can be made to be independent of π. We give three examples here; a fourth is provided by the work in [4].

<u>(a)</u> Let $\mathcal{G}$ be the smallest class of nilpotent groups with the following properties:

(1) if G is Abelian, then $G \in \mathcal{G}$.

(2) Suppose that G contains a subgroup G_0 with the following properties:

(i) G_0 contains a central subgroup Z such that $G/Z \in \mathcal{G}$;

(ii) every irreducible representation in general position for G_0 induces to an irreducible of G;

(iii) every irreducible $\pi \in G^\wedge$ in general position is induced from a representation π_0 of G_0 which is trivial on Z.

Then $G \in \mathcal{G}$.

The Heisenberg groups are all members of $\mathcal{G}$; more generally, the OKP groups of [11] are all in $\mathcal{G}$.

Theorem 4.1 Let (G, H, K) be a multiplicity-free triple with H/K Abelian such that $G \in \mathcal{G}$ and such that ρ contains representations of G in general position. Then $\lambda \times \rho$ has a strong orbital decomposition.

Proof We use induction on dim G, following the analysis in the proof of Theorem 3.1. Let Y span the Lie algebra z of Z, and let $X \in \mathfrak{g} \setminus \mathfrak{g}_0$. We can now go from $\mathfrak{g}_0$ to $\mathfrak{g}$ by looking at Cases 4, 5, and 6. Let s_0 be the cross-section in $\mathfrak{g}_0$ for $\bar{\mathfrak{h}} \setminus \bar{\mathfrak{g}}_0$ (where $\bar{\mathfrak{g}}_0 = \mathfrak{g}_0/z$, etc.). In Cases 4 and 5, $s = s_0 + \mathbb{R}X$ is a cross-section for $\mathfrak{h} \setminus \mathfrak{g}$; note that $z \subseteq \mathfrak{h}$. In Case 6, let s be the pre-image of s_0 under the projection from $\bar{\mathfrak{g}}_0$ to $\bar{\mathfrak{g}}$; then s is a cross-section for $\mathfrak{h}\backslash\mathfrak{g} \cong \mathfrak{h} \cap \mathfrak{g}_0\backslash\mathfrak{g}_0$. In each case, the fact that this choice of s meets the conditions of the theorem comes from the proof of Theorem 3.1.

Some examples where Theorem 4.1 applies are given in Section 6.

(b) Another class of cases is given by:

Theorem 4.2 If G is a 2-step nilpotent Lie group and (G, H, K) is multiplicity-free, then (K\G, H) has a strong orbital decomposition.

Proof Note that case 6(c) need never arise, since all noncentral elements are in $\mathfrak{g}^{(2)}$. Let π (over σ) appear in the decomposition, and let $\ell_0 \in \mathcal{O}_\pi$. Suppose that s is any cross section for $\mathfrak{h}\backslash\mathfrak{g}$. then, since π and σ are square-integrable mod their kernels, we have

$$\mathcal{O}_\pi = r_{\ell_0} + \ell_0 \quad (r_{\ell_0} = \text{radical of } l_0),$$

$$P^{-1}\mathcal{O}_\sigma = r'_{\ell_0} \quad (= \text{radical of } \ell_0|_h); \text{ thus}$$

$$\int_{\mathcal{O}'_{\pi,\sigma}} (f\circ\beta)(\ell)d\ell$$

$$= \int_{(r'_{\ell_0}+r_{\ell_0}+\mathfrak{k})} (f\circ\beta)^\wedge(\ell_0+\ell)d\ell.$$

Let $W \in \mathfrak{h}$ and $Y \in s$. Then

$$\beta(W, Y) = \exp W \exp Y = \exp(W + Y - \tfrac{1}{2}[W, Y]).$$

But [W, Y] is central, and hence in r_{ℓ_0}; thus $\ell([W, Y]) = 0$ for all $\ell \in (r'_{\ell_0} + r'_{\ell_0} + \mathfrak{k})$. So, by a change of variable,

$$\int_{\mathcal{O}'_{\pi,\sigma}} (f \circ \beta)^{\wedge}(\ell) d\ell$$

$$= \int_{\mathcal{O}'_{\pi,\sigma}} \left(\int_{\mathfrak{k} \setminus \mathfrak{g}} (f \circ \exp)(Y) e^{-2\pi i(\ell_0 + \ell)(Y)} dY d\ell, \right.$$

which is independent of β. In particular, we can make β independent of π.

<u>(c)</u> Symmetric space for nilpotent groups. (The results given here were obtained in [1] by different methods.) Let G have an involutive automorphicm ω, and let $K = \{x \in G : \omega(x) = x\}$, $S = \{x \in G : \omega(x) = x^{-1}\}$. Denote by $\mathfrak{k}$, s respectively the +1, −1 eigenvalues in $\mathfrak{g}$.

The following elementary result is the key to much of what follows. Recall that a characteristic subgroup (or subalgebra) of G (or $\mathfrak{g}$) is one that is invariant under all automorphism.

Lemma 4.1 With the above notation,

(a) If C is a connected characteristic Lie subgroup of G, then $C = (C \cap K) \cdot (C \cap S)$, similarly, if $\mathfrak{c}$ is a characteristic subalgebra, then $\mathfrak{c} = (\mathfrak{c} \cap \mathfrak{k}) \oplus (\mathfrak{c} \cap s)$.

(b) If H is a connected Lie subgroup of G normalizing K, with $H \supseteq K$, then $H = K \times (H \cap S)$ (direct product of groups); similarly, if $\mathfrak{k} \subseteq \mathfrak{h} \subseteq \mathfrak{g}$ and $\mathfrak{h}$ normalizes $\mathfrak{k}$, then $\mathfrak{h} = \mathfrak{k} \times (\mathfrak{h} \cap s)$ (direct product of algebras).

Proof It suffices to deal with the algebra part (one could give similar proofs for the group case anyway).

(a) This is trivial. If $X \in \mathfrak{c}$, then $2X = (X + \omega(X)) + (X - \omega(X))$, and $\omega(X) \in \mathfrak{c}$, $X + \omega(X) \in \mathfrak{k}$, $X - \omega(X) \in s$.

(b) This is nearly as trivial. We need to know that if $X \in$

$\mathfrak{h} \cap s$ and $Y \in \mathfrak{k}$, then $[X, Y] = 0$. But $[X, Y] \in \mathfrak{k}$, since $\mathfrak{h}$ normalizes $\mathfrak{k}$, and $[X, Y] \in s$, since $[\mathfrak{k}, s] \in s$ (just apply ω).

It is shown in [5] (and, of course, in [1]) that $(K\backslash G, K)$ is already multiplicity-free. We now prove the following result:

Theorem 4.3 Let H_1 be any central subgroup in G contained in S; then $(K\backslash G, KH_1)$ has a strong orbital decomposition given by s, with $\beta_0 = \exp$.

Note Let $H = KH_1$. Then $\mathfrak{k}\backslash\mathfrak{h}$ is naturally a subspace s_0 of s. Let s_1 be any complementary subspace. If $W_j \in s_j$ ($j = 1, 2$), then $\exp W_1 \exp W_2 = \exp(W_1 + W_2)$, since W_1 is central. Thus in this case β amounts to the exponential map on s.

Proof We proceed by induction, following the cases in the proof of Theorem 3.1. Just as in that proof, we assume that we have a rationally varying family of nilpotent Lie algebra; we assume further that ω is a fixed map on $\mathbb{R}^n$ (so that the subspaces corresponding to $\mathfrak{k}$ and s are independent of α). Note that every $\sigma \in K^\wedge$ is 1-dimensional.

<u>Case</u> <u>1</u>: this causes no problem.

<u>Case</u> <u>2</u>: from Lemma 4.1 and Case 1, we may assume that the central subgroup Z is in $\mathfrak{h} \cap s$. This case now follows from the case of $(Z\backslash KZ, Z\backslash H, Z\backslash G)$.

<u>Case</u> <u>3</u>: again, we may assume that the central subgroup Z in in s; the reduction to Case 2 now causes no trouble.

<u>Case</u> <u>4</u>: note that $\mathfrak{k} = \mathfrak{k}_0$, so that $\mathfrak{g}_0 = \mathfrak{k} \oplus s_0$, $s = \mathfrak{g}_0 \cap s$. We merely need to pick $X \in s$, and the proof of Case 4 in Theorem 3.1 shows that if s_0 is a cross-section for (K, H_0, G_0), then s is one for (K, H, G). Note that here we can take $\beta_0 = \exp$.

<u>Case</u> <u>5</u>: this case cannot arise. For suppose that $z^{(2)} \cap \mathfrak{k}$

$= (0)$, but that $Y \in (z^{(2)} \setminus z) \cap \mathfrak{h}$. Then (Lemma 4.1) we may assume that $Y \in \mathfrak{k}$ or $Y \in \mathfrak{s}$. We have eliminated the first possibility by assumption; hence $Y \in \mathfrak{s}$. Let Z_0 be central (and nonzero), and let $[X, Y] = Z_0$. If $X = X_1 + X_2$, $X_1 \in \mathfrak{k}$ and $X_2 \in \mathfrak{s}$, then $[X_2, Y] \in \mathfrak{k}$ is central; hence $[X_2, Y] = 0$. Thus $[X_1, Y] \neq 0$, $X_1 \in \mathfrak{k}$, and $Y \in \mathfrak{s}$; this violates Lemma 4.1(b).

Case 6: Subcase 6(b) and 6(c) cannot arise, since they require $X, Y, Z \in \mathfrak{s}$ with $[X,Y] = Z$, while $[\mathfrak{s},\mathfrak{s}] \subseteq \mathfrak{k}$. Subcase 6a is handled in the same way as Case 4. This completes the proof.

§5.

In this section, we prove the following result:

Theorem 5.1 Let (G, H, K) be a multiplicity-free triple, and let L be a (nonzero) differential operator commuting with $\lambda \times \rho$ on $\mathcal{L}^2(K\backslash G)$. Suppose that $(K\backslash G, H)$ has a strong orbital decomposition. Then L has a tempered fundamental solution; i.e., there is a tempered distribution ξ on $K\backslash G$ such that $L\xi = \delta_e$ (the delta distribution on Ke).

Notes

1. Suppose that $f \in \mathcal{S}(K\backslash G)$. Then one can find $\phi \in \mathcal{S}(G)$ such that $f = \rho(\phi)\delta_e$ (note that $\rho(\phi)\delta_e(x) = \int_K \phi(kx)dk$; if $u = \rho(\phi)\xi$ (where $(\rho(\phi)\xi)(x) = (\xi_y, \phi(x^{-1}y))$, then $Lu = f$. So L is, in particular, locally solvable.

2. In the symmetric space (§4c), the corresponding theorem, with H = K, was proved in [1].

The proof of Theorem 5.1 depends on the following lemma:

Lemma 5.2 In the situation of Theorem 5.1, parametrize the generic representation $\sigma \times \pi$ appearing in the decomposition of $\lambda \times \rho$ by the points in a Zariski-open set of an algebraic variety V (so that $\sigma \times \pi = \sigma_{\ell 0} \times \pi_\ell$; see §1). Then $(\sigma \times \pi)(L) = f_L(\sigma, \pi)I$, where f_L is a polynomial.

Proof From, e.g., Proposition 1.2 of [12], L may be regarded as an element of $\mathfrak{u}(\mathfrak{g})$ that commutes with $\mathfrak{k}$ mod the ideal $\mathfrak{u}(\mathfrak{g})\mathfrak{k}$. Now regard L as acting (on the left) on $\mathcal{L}^2(K\backslash H \times G)$. Under the direct integral decomposition of $\mathcal{L}^2(H \times G)$ given by the Plancherel theorem, the action of L on the space corresponding to σ

$\times\ \pi$ is that of a differential operator with polynomial coordinates, these coordinates depending polynomially on $\sigma \times \pi$. The same applies if we consider the decomposition of $\mathcal{S}((K\backslash H) \times G)$ or of $\mathcal{S}'(H \times G) \subseteq \mathcal{H}^{-\infty}(H \times G)$. Regard $\mathcal{L}^2(K\backslash G)$ as consisting of functions on $(K\backslash H) \times G$ with $f(Kh, g) = f(1, h^{-1}g)$ for all $h \in H$, $k \in K$, $g \in G$. Then $\mathcal{S}'(K\backslash G)$ is decomposed under the decomposition of $\mathcal{S}'((K\backslash H) \times G)$, and the action of L on the space corresponding to $\sigma \times \pi$ is a differential operator with polynomial coefficients depending polynomially on $\sigma \times \pi$. But since $\lambda \times \rho$ is multiplicity-free, and L is central, L is scalar on $\sigma \times \pi$. The lemma follows.

We now prove Theorem 5.1; the proof is essentially that of [19] (or [1]). Let f_{L^t} be the polynomial corresponding to L^t; note that if L commutes with $\sigma \times \rho$, so do L^t, and L^*. Considering $(L^*L)^t$ instead of L^t, we may assume that f_{L^t} is nonnegative. Let $P(\ell) = f_L(\pi \times \rho)$ if $\ell \in \mathcal{O}_\pi$ and $\ell|_h \in \mathcal{O}_\sigma$. Identify $K\backslash G$ with $(\mathfrak{k}\backslash\mathfrak{h}) \times \mathcal{S}$ via β, and identify $(\mathfrak{k}\backslash\mathfrak{h}) \times \mathcal{S}$ with $\mathfrak{k}\backslash\mathfrak{g}$ in the obvious way. Define a distribution P^s on $(\mathfrak{k}\backslash\mathfrak{g})^* = k^\perp$ by

$$(P^s, \phi) = \int_{\mathfrak{k}} P^s(\ell)\phi(\ell)d\ell, \qquad \phi \in \mathcal{S}(k^\perp),$$

and define $L(s)$ on $\mathfrak{k}\backslash\mathfrak{g}$ by $(L(s), f) = (P^s, f^\wedge)$. Then

$$(L(s + 1), f) = (P^{s+1}, f^\wedge) = \int_{\mathfrak{k}^\perp} P^{s+1}(\ell)f^\wedge(\ell)d\ell$$

$$= \int_V \int_{\mathcal{O}'_{\pi,\sigma}} P^{s+1}(\ell)f^\wedge(\ell)d_{\sigma,\pi}(\ell)dn(\sigma, \pi)$$

(V = space of generic σ, π)

$$= \int_V P^{s+1}(\ell)\mathrm{Tr}(Af)(\sigma, \pi)(e)d\nu(\sigma, \pi)$$

$$= \int_V P^s(\ell)\mathrm{Tr}(A(L^t f)(\sigma, \pi)(e) d\nu(\sigma, \pi)$$

$$= \int_V \int_{\mathcal{O}'_{\pi,\sigma}} P^s(\ell)(L^t f)(\ell) d_{\sigma,\pi}(\ell) d\nu(\sigma, \pi)$$

$$= \int_{\mathfrak{k}^L} P^s(\ell)(L^t f^\frown)(\ell) d\ell = (L(s), L^t f),$$

or $L(s + 1) = LL(s)$ for $\mathrm{Re}(s) > 0$. Now P^s has a meromorphic continuation to all $s \in \mathbb{C}$; correspondingly, $L(s)$ has a meromorphic continuation, with $L(s + 1) = LL(s)$ and $L(0) = \delta_e$. Let

$$L(s) = \sum_{j=-m}^{\infty} \xi_j (s + 1)^j$$

be the Laurent series for $L(s)$ about -1. Then

$$LL(s) = L(s + 1) = \sum_{j=-m}^{\infty} L^t \xi_j (s + 1)^j \quad \text{near } s = -1;$$

writing s for $s + 1$, we see that $L^t \xi_0 = \delta$. Thus L has a tempered fundamental solution, as claimed.

§6.

We collect here some examples and open problems.

1. Let G be the solvable (non-nilpotent) real Lie group with Lie algebra $\mathfrak{g}$ spanned by W, X, Y, Z, satisfying

$$[W, Y] = Y, \ [W, Y] = -X, \ [X, Y] = Z, \ Z \text{ central};$$

let $K = \exp \mathbb{R}W$, $\mathfrak{n} = \operatorname{span}(X, Y, Z)$. Then the elements of $D(K\backslash G)$ are given by those elements of $\mathfrak{u}(\mathfrak{n})$ commuting with W; some calculation shows that these are the elements of $\mathbb{R}[Z, X^2 + Y^2]$. Thus $\mathbb{D}(K\backslash G)$ is commutative. But $X^2 + Y^2 + iZ$ is not locally solvable; i.e., there are elements of $\mathbb{D}(K\backslash G)$ which are not locally solvable. (This example, due to R. Lipsman, is the one mentioned in §1.) For this G, the normalizer of $\mathfrak{k}$ (= span W)_ is $\mathfrak{h} = \operatorname{span}(W, Z)$, and (G, H, K) is not multiplicity-free.

2. In the nilpotent case, too, there exist pairs (G, K) such that if H is the normalizer of K, then (G, H, K) is not multipicity-free. For one example, let G be the 11 dimensional Lie group whose Lie algebra $\mathfrak{g}$ is spanned by $X_1, \ldots, X_{11}$, with

$$\sum_{j=1}^{11} a_j X_j \longleftrightarrow \begin{bmatrix} 0 & a_1 & a_5 & a_7 & a_3 & a_9 & a_{11} \\ 0 & 0 & a_3 & a_4 & 0 & 0 & a_{10} \\ 0 & 0 & 0 & 0 & 0 & a_3 & a_8 \\ 0 & 0 & 0 & 0 & 0 & a_4 & a_6 \\ 0 & 0 & 0 & 0 & 0 & 0 & a_4 \\ 0 & 0 & 0 & 0 & 0 & 0 & a_2 \\ 0 & 0 & 0 & 0 & 0 & 0 & 0 \end{bmatrix}$$

Incidentally, G has square-integrable representations, but the functionals corresponding to these representations do not have any polarizing subalgebras that are ideals. Let $\mathfrak{k}$ be spanned by X_1. The algebra $\mathfrak{h}$ normalizing $\mathfrak{k}$ is spanned by X_1, X_2, X_5, X_6, X_7, X_8,

X_9, and X_{11}; the center of $\mathfrak{k}\backslash\mathfrak{h}$ is X_{11} (or its image mod $\mathfrak{k}$), $K\backslash H$ has square-integrable representations, and each of these induces to an infinite multiple of the representation of G with the same central character. Incidentally, $Z\mathbb{D}(K\backslash G) = \mathbb{C}[X_{11}]$, so that we have local solvability for $Z\mathbb{D}(K\backslash G)$.

3. On the other hand, let G be the group of upper 4×4 triangular matrices; its Lie algebra $\mathfrak{g}$ is spanned by the E_{ij} $(1 \leqslant i < j \leqslant 4)$, where the entry in the $(k\ell)^{th}$ place of E_{ij} is $\delta_{ik}\delta_{j\ell}$. Let $\mathfrak{k}$ be spanned by E_{12}; then we can let $\mathfrak{h}$ be spanned by E_{12}, E_{13} and E_{14}, and E_{34}; then $(K\backslash G, H)$ also has a strong orbital decomposition, as one discovers by applying the proof of Theorem 1.1 to this situation.

4. Let G be the 4-dimensional group whose Lie algebra $\mathfrak{g}$ is spanned by W, X, Y, Z, with

$$[W, X] = Y, \ [W, Y] = Z, \ [X, Y] = Z, \ Z \text{ central}.$$

The theory now applies to any closed connected subgroup K not meeting the center, if H is the normalizer of K. A number of cases were dealt with in [4]; the case where $\mathfrak{k}$ = span W, which does not yield to the methods of [4], was one of the notivations for this paper.

Finally, here are some questions suggested by the results given here.

1. Is thre always an orbital decomposition in the multiplicity-free case? I strongly suspect so; the ideas in [23] may be of help in proving this or a similar result.

2. Is there always a strong orbital decomposition in the multiplicity-free case? If so, what properties characterize the allowable cross-section s? If not, what properties characterize the groups with strong orbital decomposition?

3. Is there a case of a nilpotent group G with a connected

subgroup K such that $Z\mathbb{D}(K\backslash G)$ contains an unsolvable operator?

4. To what extent can one obtain this sort of orbital picture for exponential solvable groups, or for general solvable groups?

REFERENCES

1. Benoist, Y., Espaces Symmétriques Exponentiels, Thèse 3ème Cycle, Univ. Paris VII (1983).

2. Corwin, L., A representation-theoretic criterion for local solvability of left invariant operators on nilpotent Lie groups, Trans. Am. Math. Soc. 264 (1981), 113-120.

3. Corwin, L., Solvability of Left Invariant Operators on Nilpotent Lie Groups, in Lie Group Representations III, Springer Lecture Notes in Mathematics, 1077 (1984).

4. Corwin, L., and Greenleaf, F. P., Harmonic analysis on certain homogeneous spaces of nilpotent groups, preprint.

5. Corwin, L. and Greenleaf, F. P., Direct integral decompositions and multiplicities for induced representations of nilpotent Lie groups, preprint.

6. De George, D., and Wallach, N., Limit formulas for multiplicities in $L^2(\Gamma\backslash G)$, Annals of Math. 107 (1978), 133-150.

7. Duflo, M., Opérateurs différentiels bi-invariants sur un groupe de Lie., Ann. Sci. Ec. Norm. Sup. 10 (1977), 265-286.

8. Grelaud, G., Sur les représentations des groupes de Lie résolubles, Thèse, Univ. de Poitiers (1984).

9. Helgason, S., Invariant differential operators and eigenspace representations, in M. Atiyah et al., Representation Theory of Lie Groups, Cambridge, Cambridge University Press (1979), 236-286.

10. Hormander, L., Differential equations without solutions, Math. Annalen 140 (1960), 169-173.

11. Howe, R., Ratcliff, G., and Wildberger, N., Symbol mappings for certain nilpotent groups, preprint.

12. Jacobson, J., Invariant differential operators on some homogeneous spaces for solvable Lie groups, Aarhus Universitet, Matematisk Insitut, Various Publications Series No. 34 (1982).

13. Kirillov, A. A., Unitary representations of nilpotent Lie groups, Uspehi Mat. Nauk 17 (1962), 57-110; translation in Russian Math. Surveys 17 (1962), 53-104.

14. Lewy, H., An example of a smooth linear partial differential equation without solution, Ann. Math. 66 (1957), 155-158.

15. Lion, G., Hypoellipticité et résolubilité d'opérateurs différentiels sur des groupes nilpotents de rang 2, Comptes Rendus Acad. Sci. (Paris) 290 (1980), Ser. A., 271-274.

16. Mackey, G., The Theory of Unitary Group Representations, Chicago, University of Chicago Press. 1976.

17. Metiver, G., Equations aux derivées partielles sur les groupes de Lie nilpotents, Expose #538, Seminaire Bourbaki 1981-82.

18. Pukanszky, L., Unitary representations of solvable Lie groups, Ann. Sci. Ec. Norm. Sup. 4(4) (1971), 457-608.

19. Raïs, M., Solutions élémentaires des opérateurs différentiels bi-invariants sur un groupe de Lie nilpotent, Comptes Rendus Acad. Sci. (Paris) 273 (1971), Ser. A., 495-498.

20. Rothschild, L., Local solvability of left-invariant operators on

the Heisenberg group, Proc. Am. Math. Soc. 74 (1979), 383-388.

21. Rouvière, F., Sur la résolubilité locale des opérateurs bi-invariants, Annali Scuola Normale Superiore (Pisa) 3 (1976), 231-244.

22. Segal, I., The two-sided regular representation of a unimodular locally compact group, Annals of Math. 51 (1950), 293-298.

23. Wildberger, N., Quantization and Harmonic Analysis on Nilpotent Lie Groups, Thesis, Yale University, 1983.

24. Bernat, P., et al, Représentations des groupes de Lie résolubles. Paris, Dunod, 1972.

25. Corwin, L., Criteria for solvability of left invariant operators on nilpotent Lie groups, Transactions Am. Math. Soc. 280 (1983), pp. 53-72.

NOTE The results in [5] and [8] will be published in a joint paper by all three authors.

SOME HOMOTOPY AND SHAPE CALCULATIONS FOR C*-ALGEBRAS

Edward G. Effros
Jerome Kaminker

Dedicated to George Mackey, who inspired a generation of students with his vision of the unity of mathematics.

1. INTRODUCTION.

Suppose that a C^*-algebra A is isomorphic to the direct limit $\varinjlim A_n$ of a system of C^*-algebra and injections

(1.1) $$A_1 \xrightarrow{\varphi_1} A_2 \xrightarrow{\varphi_2} \ldots\ldots$$

The goal of shape theory [EK1] is to extract invariants for the C^*-algebra A from the <u>homotopy diagram</u>

(1.2) $$A_1 \xrightarrow{[\varphi_1]} A_2 \xrightarrow{[\varphi_2]} \ldots\ldots$$

where $[\varphi_n]$ denotes the homotopy class of φ_n. There are natural categories of such diagrams (see §5), and under favorable circumstances, any two systems (1.1) with limits isomorphic to A must determine isomorphic diagrams (1.2). In that case the uniquely determined isomorphism class of (1.2) is called the <u>shape</u> of A.

Turning to the simplest example, suppose that the A_n are finite dimensional and the φ_n are unital. Then A is an AF (approximately finite) algebra, and the homotopy diagram (1.2) is known as a <u>Bratteli diagram</u> for A. Generalizing results of Glimm [G1] and Dixmier [D1], Bratteli proved in [B1, Th. 2.7] that any two homotopy diagrams for a given AF algebra A must be isomorphic. He went on to show that shape provides a complete invariant for AF algebras. Subsequently, Elliott [E1] introduced an algebraic model for the shape invariant of an AF algebra. Since the homotopy class $[\varphi_n]$ is determined by the corresponding map

$$\varphi_{n*} = K_0(\varphi_n):K_0(A_n) \longrightarrow K_0(A_{n+1}),$$

one may replace (1.2) by a system of ordered groups with distinguished "units" (determined by the C*-algebraic identities), and unital positive homomorphisms

$$K_0(A_1) \xrightarrow{\varphi_1{}^*} K_0(A_2) \xrightarrow{\varphi_2{}^*} \cdots\cdots \tag{1.3}$$

These group diagrams may in turn be replaced by their unital ordered limits $K_0(A) = \varinjlim K_0(A_n)$. One thus has that the following are equivalent for AF C^*-algebras A and B:

(1) A is isomorphic to B,

(2) A and B have the same shape (i.e., isomorphic homotopy diagrams),

(3) $K_0(A) \cong K_0(B)$ (as ordered, unital groups).

The applicability of the AF-theory is limited by the fact that such algebras are in some sense zero-dimensional. In an effort to find "higher dimensional" analogues, it is natural to consider systems of Cuntz-Krieger algebras

$$0_{\underline{a}_1} \longrightarrow 0_{\underline{a}_2} \longrightarrow \cdots\cdots \tag{1.4}$$

On the one hand, one may view the $0_{\underline{a}}$'s as being one-dimensional, since they can usually be described as corners of cross-products

$\mathbb{Z} \times_\alpha A$ with A an AF-algebra (recall that if α is trivial, $\mathbb{Z} \times_\alpha A$ = $C(S^1) \otimes A$). On the other hand, they are especially amenable to analysis since one can completely describe the collection of homotopy classes of the connecting maps ([C1]-see §4).

It is useful to consider three cases of such limits (1.4):

a) the matrices $\underline{a}_n$ are irreducible, but not cyclic permutations (and thus the $O_{\underline{a}}$ are simple), and the φ_n are <u>proper</u>, i.e., $0 < \varphi_n(1) < 1$,

b) the $\underline{a}_n$ are non-cyclic irreducible, and $\varphi_n(1) = 1$,

c) the a_n are cyclic permutation matrices (and thus $O_{\underline{a}_n} = C(S^1) \otimes M_p$), and $\varphi_n(1) = 1$.

It follows from [EK1] that in cases (b) and (c), the isomorphism class of the homotopy diagram

$$O_{\underline{a}_1} \longrightarrow O_{\underline{a}_2} \longrightarrow \cdots\cdots$$

is uniquely determined by the limit $\varinjlim O_{\underline{a}_n}$. Those arguments extend to yield this result for (a) as well (see §6). Thus, if A = $-O_{\underline{a}_n}$ the shape of A is well-defined.

In this paper we restrict our attention to case (a), and we say that a C^*-algebra A is an <u>AO_p algebra</u> if it is isomorphic to the limit of such a system. Our main result (Theorem 6.1) is that if A and B are AO_p algebras, then the following are equivalent:

(1) A and B have the same shape,

(2) $K_0(A) \cong K_0(B)$ and $K_1(A) \cong K_1(B)$ (as unordered groups),

The unital situation is more involved since the collection $[O_{\underline{a}}, O_{\underline{b}}]$ of homotopy classes of maps $\varphi: O_{\underline{a}} \longrightarrow O_{\underline{b}}$ need not form a group in a natural way. We will consider cases (b) and (c) in a subsequent paper.

As one might expect, this result is rather more difficult to prove than its AF analogue. It is also somewhat surprising since the class $[\varphi_n]$ is not completely determined by $K_0(\varphi_n)$ and $K_1(\varphi_n)$. As Cuntz showed in [C1], one must instead use the non-independent maps $K_0(\varphi_n)$ and $\mathrm{Ext}(\varphi_n)$. The latter maps initially determine a projective system of groups. Such "pro-groups" are difficult to handle since the projective limit functor is often degenerate. Fortunately, we have found a "splitting" result for pro-groups that permits one to circumvent these difficulties (see Theorem 5.2).

Central to the entire theory is Cuntz's formulas for the homotopy classes of maps $\varphi: O_{\underline{a}} \longrightarrow O_{\underline{b}}$, which were essentially given in [C1]. These results will undoubtedly play a central role in C^*-algebraic topology. We have therefore decided to include a fairly complete discussion of Cuntz's calculations in §§ 3 - 5.

We are indebted to J. Cuntz for pointing out to us the relevance of [C1] in the non-stable case, and for reassuring us about some of the needed arguments.

2. SOME K AND EXT RESULTS.

We follow Cuntz's approach to K-theory [C2]. For a fixed separable Hilbert space H, we let $\mathcal{B} = \mathcal{B}(H)$, and $\mathcal{K} = \mathcal{K}(H)$ be the bounded and compact operators, respectively, on H. For any C^*-algebra A, we let Proj A be the projections in A, and $\mathrm{Proj}_p A$ be the proper projections, i.e., those satisfying $0 < e < 1$. We say that $e, f \in$ Proj A are equivalent if there exists an $s \in A$ with $s^*s = e$, $ss^* = f$, in which case we write $s: e \simeq f$, or simply $e \simeq f$, and we let $[e]$ be the equivalence class of e. If A is unital, we let U(A) be the unitaries in A, and $U_0(A)$ be the path component of 1. We say that u and v are equivalent, and write $u \sim v$, if $uv^{-1} \in U_0(A)$. For any C^*-algebra A, we let $\tilde{A} = A \oplus \mathbb{C}1$ be the unital extension of A, and $\chi:\tilde{A} \longrightarrow \mathbb{C}$ the corresponding character $\chi(a+\alpha 1) = \alpha$.

Given a unital C^*-algebra A we say that $e,f \in \text{Proj } A\otimes K$ are K_0-equivalent, and write $[e]_0 = [f]_0$, if there is a projection $g \in \text{Proj } A\otimes K$ with $e \perp g$ and $f \perp g$, and $e+g \simeq f+g$. The set $K_0(A)$ of such classes is a group under the operation

$$[e]_0 + [f]_0 = [e' + f']_0$$

where $e \simeq e' \perp f' \simeq f$. K_0 may be regarded as a functor on the category of unital C^*-algebras and (not necessarily unital) homomorphisms since given $\varphi:A \longrightarrow B$ we may let

$$K_0(\varphi)[e]_0 = [(\varphi\otimes \mathrm{id})(e)]_0.$$

To obtain a functor on all C^*-algebras, we define $K_0'(A)$ for general A to be the kernel of the map

$$K_0(\chi):K_0(\tilde{A}) \longrightarrow K_0(\mathbb{C}).$$

Given $\varphi:A \longrightarrow B$, we let $\tilde{\varphi}:\tilde{A} \longrightarrow \tilde{B}$ be the unital extension, and we let $K_0'(\varphi)$ be the restriction of $K_0(\tilde{\varphi})$ to $K_0'(A)$. For unital algebras A, $\tilde{A} = A\oplus\mathbb{C}1$, and we have a natural isomorphism $K_0'(A) \cong K_0(A)$ under which $K_0'(\varphi)$ corresponds to $K_0(\varphi)$. Thus our convention will be to write $K_0(A) = K_0'(A)$ and $K_0(\varphi) = K_0'(\varphi)$ for arbitrary C^*-algebras A.

If A is unital we define

$$K_1(A) = U((A\otimes\mathcal{K})^{\sim})/U_0((A\otimes\mathcal{K})^{\sim}).$$

and we denote the corresponding equivalence classes by [u], (u $\in$ $U((A\otimes\mathcal{K})^{\sim})$. We define $K_1(\varphi)$ for unital homomorphisms φ:A $\longrightarrow$ B by $K_1(\varphi)([u]) = [(\varphi\otimes id)^{\sim}(u)]$. We obtain a functor on the category of all C^*-algebras and homomorphisms by letting $K_1'(A) = K_1(\tilde{A})$ and $K_1'(\varphi) = K_1(\tilde{\varphi})$. If A is unital, $K_1(\mathfrak{X}) = 0$, hence there is no problem in using the notation $K_1(A) = K_1'(A)$ and $K_1(\varphi) = K_1'(\varphi)$ for arbitrary C^*-algebras A and maps φ.

The following useful result is a special case of [P1] (we are indebted to J. Mingo for this reference). We recall that a projection e in a C^*-algebra A is <u>full</u> if the closed two-sided ideal it generates is all of A.

<u>Lemma</u> <u>2.1</u> If e is a full projection in a separable C^*-algebra A and j:eAe $\longrightarrow$ A is the inclusion map, then the induced homomorphisms $K_0(j)$ and $K_1(j)$ are isomorphisms.

Proof From Brown's Stability Theorem [B2], there exists a partial isometry v $\in$ $M(A\otimes\mathcal{K})$ such that $v^*v = 1\otimes 1$ and $vv^* = e\otimes 1$ (M denotes the double multipliers). The map

$$J: eAe\otimes\mathcal{K} \longrightarrow A\otimes\mathcal{K}$$

defined by $J(b) = v^*bv$ is an isomorphism. and thus the same is true for the maps $K_i(J)$ (i=1,2). On the other hand the commutative diagram

$$\begin{array}{ccc} eAe & \xrightarrow{j} & A \\ \downarrow & & \downarrow \\ eAe\otimes K & \xrightarrow{j\otimes id} & A\otimes K \end{array}$$

induces a commutative diagram with bijective columns

$$\begin{array}{ccc} K_*(eAe) & \longrightarrow & K_*(A) \\ \downarrow & & \downarrow \\ K_*(eAe\otimes K) & \longrightarrow & K_*(A\otimes K) \end{array}$$

and thus it suffices to show $K_*(j\otimes id) = K_*(J)$. Since $eAe\otimes K$ is stable, it suffices to consider projections f in $eAe\otimes K$ and unitaries u in $(eAe\otimes K)^{\sim}$.

Given $f \in Proj(eAe\otimes K)$, we set $v_1 = fv \in A\otimes K$. Thus in $A\otimes K$ we have

$$J(f) = v^*fv = v_1{}^*v_1 \cong v_1v_1{}^* = f = (j\otimes id)(f)$$

hence $K_0(J)[f]_0 = K_0(j\otimes id)[f]_0$. Given a unitary $u \in (eAe\otimes K)^{\sim}$ it suffices to show that u can be path connected to v^*uv in the group G of invertibles in $(A\otimes K)^{\sim}$, since polar decomposition will provide a path of unitaries. Letting G_0 be the component of the identity in G, G_0 is open. We have $x,y \in G$ can be path-connected if the only if $xy^{-1} \in G_0$, in which case we write $x\sim y$. We see that if x and $y \in G$ are sufficiently close, then $x\sim y$. We have $u = c' + \alpha 1\sim c + 1$, where $|\alpha| = 1$ and c' and $c = \alpha^{-1}c'$ are normal in $eAe\otimes K$. Choosing d from an approximate identity in the commutative C^*-algebra generated by c, we may assume that $cd + 1$, $v^*cdv + 1 \in G$ and

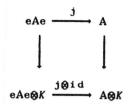

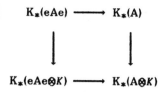

$$c \oplus 1 \sim cd \oplus 1 = dc \oplus 1.$$

hence,

$$v^*cv \oplus 1 \sim v^*cdv \oplus 1.$$

Letting $a = v^*c$, $b = dv$, we have

$$ba \oplus 1 = dvv^*c \oplus 1 = d(e \otimes 1)c \oplus 1 = dc \oplus 1 \sim u,$$

$$ab \oplus 1 = v^*cdv \oplus 1 \sim v^*uv.$$

Appealing to the matrix trick in [P1], $ba \oplus 1 \sim ab \oplus 1$, and thus $u \sim v^*uv$. ∎

Cuntz [C2] has made the fundamental observation that K-theory is greatly simplified if there are sufficiently many infinite projections. We will need the details of this theory in our proof of the Universal Coefficient Theorem (see Theorem 3.1).

A projection e in a C^*-algebra is <u>infinite</u> if there is a projection $e' \in A$ with $e \sim e' < e$. A collection P of proper projections in a C^*-algebra A is said to be <u>ample</u> if it satisfies the <u>Cuntz conditions</u>

(C_1) If $e, f \in P$ and $ef = 0$, then $e + f \in P$.

(C_2) If $e \in P$ and $e' \in \mathrm{Proj}_p A$ satisfies $e' \sim e$, then $e' \in P$.

(C_3) For all $e, f \in P$ there exists an $e' \in P$ such that $e \sim e' < f$ and $f - e' \in P$.

(C_4) If $e \in \mathrm{Proj}_p A$ and $e \geq f \in P$, then $e \in P$.

We note that (C_3) implies that the projections in P are infinite. If A has an ample collection of projections P, we may identify $K_0(A)$ with the equivalence classes $[e]$ ($e \in P$) together with the operation

$$[e] + [f] = [e' + f']$$

where $e', f' \in P$ satisfy $e \cong e' \perp f' \cong f$. Making some obvious modifications to provide proper projections, this is shown in [C2].

Cuntz proved that if A is a simple C^*-algebra and the collection P of infinite proper projections is non-empty, then P is ample. We shall need a slight variation on this result. Given a unital extension E of a unital C^*-algebra A of the form

$$(2.1) \qquad 0 \longrightarrow K \longrightarrow E \xrightarrow{\pi} A \longrightarrow 0,$$

we say that $e \in \text{Proj}\, E$ is <u>non-compactly</u> <u>infinite</u> if $e \sim e' < e$ where $e - e' \notin K$. The extension (2.1) is said to be <u>essential</u> if $0 \neq a \in E$ implies that $ax \neq 0$ for some $x \in K$.

<u>Lemma</u> <u>2.2</u> Given a unital essential extension (2.1) where A is unital, simple, and has an infinite projection, the collection P of proper non-compactly infinite projections in E is ample.

Proof The sequence (2.1) determines an injection $\theta: E \longrightarrow M(K) = \mathbb{B}$. We may identify E with its image in $\mathbb{B}$ and π with the restriction of the quotient map $\pi: \mathbb{B} \longrightarrow Q$. Given $s: e \simeq e' < e$ in A, we have that $v = s + (1 - e)$ satisfies $v: 1 \simeq f = e' + (1 - e) < 1$. We may lift v to an isometry $v' \in \mathbb{B}$ as folllows. Choosing $b \in \mathbb{B}$ with $\pi(b) = v$, we have $\pi(b^*b) = 1$, and thus $\pi(|b|) = 1$. Taking the polar decomposition $b = w|b|$, we have w is a partial isometry with $\pi(w) = \pi(b) = v$. Since $\pi(w^*w) = 1$ (resp., $\pi(ww^*) < 1$), w^*w (resp., ww^*) is a projection of finite (resp., infinite) codimension. We may thus let $v' = w + u$ for an appropriate finite rank partial isometry u.

Letting $f' = v'v'^*$, we have $1 \simeq f'$ and $1 - f' \notin K$, hence 1 is non-compactly infinite. Noting that $f' \simeq v'f'v'^*$, it follows that f'

$\in P$. As in [C2], the only non-trivial condition is (C_3). If $J \neq 0$ is a (not necessarily closed) ideal in E and $J \not\subset K$, then $J = A$. To see this, note that $J \cap K \neq 0$ since if $a \in J$, $ak \neq 0$ for some $k \in K$. It follows that $\bar{J} \supseteq K$, and since A is simple, $\bar{J} = E$. We have for some $a \in J$, $\|1-a\| < 1$, and thus a is invertible in E and $1 = aa^{-1} \in J$, i.e., $J = E$. Given $e,f \in P$, we let $v:f \sim f' < f$, where $r = f - f' \notin K$. Then since the ideal generated algebraically by r is E itself, there are element $a_i.b_i \in A$ with $e = \Sigma a_i r b_i$. The remainder of the argument proceeds as in [C2, §1.5]. ■

A simple C^*-algebra A is said to be <u>purely</u> <u>infinite</u> [C2] if for each $a \in A$ with $0 \leqslant a \neq 0$, aAa contains an infinite projection. The K-Theory for such an algebra A simplifies even further. To begin with, all non-zero projections are infinite, hence $\text{Proj}_p A$ is itself ample. Furthermore, $K_1(A)$ may be identified with $U(A)/U_0(A)$ [C2]. Finally, we will prove in Corollary 2.4 that on $\text{Proj}_p A$, one may take homotopy to be the equivalence relation.

To begin with, given $e,f \in \text{Proj}_p A$ with $e \simeq f$, there is a $u \in U(A)$ such that $ueu^* = f$. To see this we first "shrink" 1 to a proper projection, since we wish to use proper projections to model $K_0(A)$. We fix an equivalence $v:1 \simeq g < 1$, and let $e' = vev^*$, $f' = vfv^*$. We have $e',f' < g$, and letting $s:e \simeq f$, it follows that $vsv^*:e' \simeq f'$. Thus

$$[e'] + [g - e'] = [g] = [f'] + [g - f']$$

and $[g - e'] = [g - f']$. Letting $w':g - e' \simeq g - f'$. It follows that $w = v^*w'v:1 - e \simeq 1 - f$, and $u = s + w$ has the desied property.

It will follow from the next result that the u given above can be chosen in $U_0(A)$.

<u>**Lemma**</u> <u>2.3</u> Suppose that A is a purely infinite simple C^*-algebra and that $e \in \text{Proj}_p A$ and $u \in U(A)$ are given. Then there

exists a unitary $u' \in U(eAe)$ such that

$$u \simeq u' + (1 - e)$$

Proof Since A is simple, Lemma 2.1 implies that $j: eAe \longrightarrow A$ induces an isomorphism

$$K_1(j): K_1(eAe) \longrightarrow K_1(A)$$

Since A and thus eAe are purely infinite, we may identify $K_1(j)$ with the map

$$U(eAe)/U_0(eAe) \longrightarrow U(A)/U_0(A)$$

induced by $u' \longrightarrow u' + (1 - e)$. Our assertion follows. ■

__Corollary 2.4__ Suppose that A is a purely infinite simple C^*-algebra and that $e,f \in \mathrm{Proj}_p A$. Then the following are equivalent:

(1) $e \simeq f$

(2) there is a unitary $u \in U_0(A)$ such that $ueu^* = f$

(3) e and f can be connected by a path in $\mathrm{Proj}_p A$.

Proof (1) ⟹ (2). Given $e \cong f$, there is a unitary $u \in U(A)$ such that $ueu^* = f$ (see above). From Lemma 2.1 there is a unitary $w \in U((1 - f)A(1 - f))$ such that $[f + w]_1 = -[u]_1$ in $K_1(A) = U(A)/U_0(A)$. We have that $u_0 = (f + w)u \in U_0(A)$, and $u_0 e u_0^* = f$.

(2) ⟹ (3) is trivial.

(3) ⟹ (1) is a simple consequence of the fact that if $e,f \in \mathrm{Proj}_p A$ satisfy $\|e - f\| < 1$, then $e \simeq f$ (see [E2]). ■

It should be noted that the above result is false in Proj A since no proper projection can be joined to either 0 or 1.

Given unital C^*-algebras A and B we say that a homomorphism $\varphi: A \longrightarrow B$ is <u>proper</u> if φ is one-one and $0 < \varphi(1) < 1$. We let $Hom_p(A,B)$ denote the set of all such maps with the point-norm topology. We say that $\varphi, \psi \in Hom_p(A,B)$ are <u>homotopic</u> if they can be connected by a continuous path of proper homomorphisms $\theta_t \in Hom_p(A,B)$ $(0 \leq t \leq 1)$, and we then write $\varphi \simeq \psi$. We let $[A,B]_p$ denote the collection of equivalence classes $[\varphi]$. If $\varphi(1) \perp \psi(1)$ and $\varphi(1) + \psi(1) < 1$, then $\varphi + \psi \in Hom_p(A,B)$. If B is purely infinite, we may define a semigroup operation on $[A,B]_p$ by

$$[\varphi] + [\psi] = [\varphi' + \psi'],$$

where $\varphi \simeq \varphi'$, $\psi \simeq \psi'$, and $\varphi'(1) \perp \psi'(1)$, $\varphi'(1) + \psi'(1) < 1$. To see that φ', ψ' exists, we may let $\varphi(1) \simeq e' \perp f' \simeq \psi(1)$, $e' + f' < 1$. From Corollary 2.4 we may choose $u,v \in U_0(B)$ with $u\varphi(1)u^* = e'$, $v\psi(1)v^* = f'$. It is evident that $\varphi' = u\varphi u^*$ and $\psi' = v\psi v^*$ have the desired properties.

Turning to particular cases, we have the homomorphisms $\varphi: \mathbb{C} \longrightarrow B$ correspond exactly to the projections $e = \varphi(1) \in B$. Thus if B is purely infinite and simple, we see that

$$[\mathbb{C},B]_p = K_0(B).$$

One the other hand, the Calkin algebra $Q = \mathcal{B}(H)/\mathcal{K}(H)$, H separable, is a purely infinite simple C^*-algebra [C2]. For any simple, separable nuclear C^*-algebra A we have that $[A,Q]_p$ is a group, and in fact

$$[A,Q]_p = Ext(A),$$

(see [K1, §7], [S2]). The parallel with Corollary 2.4 in this case is particularly striking since it follows that

(2.2) $\varphi \simeq \psi$ if and only if $u\varphi u^* = \psi$ for some $u \in U_0(Q)$.

The natural pairing between $K_1(A)$ and $Ext(A)$ and the corresponding Universal Coefficient Theorem for $A = 0_{\underline{a}}$ will play an important role in the paper. We recall that $K_0(Q) = 0$ (i.e., if $e,f \in Proj_p Q$, then $e \simeq f$), and the Fredholm index provides us with an isomorphism $ind: K_1(Q) \longrightarrow \mathbb{Z}$. Given $u \in U(Q)$, we will generally write $ind\ u = ind\ [u]$. Given a projection e in a purely infinite, simple, unital, nuclear C^*-algebra A, the isomorphism $K_1(eAe) \longrightarrow K_1(A)$ of Lemma 2.1 is implemented by $[u] \longrightarrow [\tilde{u}]$, where $\tilde{u} = u + (1 - e)$. Given $u \in U(eAe)$, we let $ind_e u = ind\ [\tilde{u}] \in \mathbb{Z}$. Given $[\alpha] \in Ext(A)$ where $\alpha: A \longrightarrow Q$ is a proper injection, and $[u] \in K_1(A)$ where $u \in U(A)$, we define

$$<[\alpha],[u]> = <\alpha,u> = ind_e \alpha(u) \in \mathbb{Z}$$

where $e = \alpha(1)$ (and thus $\alpha(u) \in U(eQe)$). It is simple matter to verify that this is a biadditive map

$$(2.3) \qquad < , >: Ext(A) \times K_1(A) \longrightarrow \mathbb{Z}$$

Following [B3], (2.3) determines a homomorphism

$$\gamma_\infty: Ext(A) \longrightarrow Hom(K_1(A),\mathbb{Z}).$$

in the obvious manner. On the other hand we will see below that there is a natural homomorphism (see [B3])

$$\kappa: \ker \gamma_\infty \longrightarrow Ext(K_0(A),\mathbb{Z})$$

where Ext denotes the usual extension bifunctor on abelian groups.

Before defining κ, we first recall the "external" definition of

Ext(G,H) for abelian groups G, H (see [HS1, §3.2]). An <u>extension</u> E of G by H is an exact sequence

$$0 \longrightarrow H \longrightarrow E \longrightarrow G \longrightarrow 0.$$

We say that another such extension

$$0 \longrightarrow H \longrightarrow E' \longrightarrow G \longrightarrow 0$$

is <u>equivalent</u> to it if there is a commutative diagram

$$\begin{array}{ccccccc} & & & E & & & \\ & & \nearrow & & \searrow & & \\ 0 \longrightarrow & H & & \downarrow & & G & \longrightarrow 0, \\ & & \searrow & & \nearrow & & \\ & & & E' & & & \end{array}$$

and we let Ext(G,H) be the set of equivalence classes [E].

Given a homomorphism φ:H $\longrightarrow$ H', we may associate with each extension [E] $\in$ Ext(G,H), [E'] = φ_*[E] $\in$ Ext(G,H') by using a pushout diagram, viz., the solid arrows in the resulting diagram

$$\begin{array}{ccccccccc} 0 & \dashrightarrow & H & \longrightarrow & E & \dashrightarrow & G & \dashrightarrow & 0 \\ & & \downarrow & & \downarrow & & \| & & \\ 0 & \dashrightarrow & H' & \longrightarrow & E' & \dashrightarrow & G & \dashrightarrow & 0. \end{array}$$

Similarly, a homomorphism ψ:G' $\longrightarrow$ G determines an extension [E'] = ψ^*[E] $\in$ Ext(G',H) by pull-back, viz., the solid arrows in the resulting diagram

$$\begin{array}{ccccccccc} 0 & \dashrightarrow & H & \dashrightarrow & E' & \longrightarrow & G' & \dashrightarrow & 0 \\ & & \| & & \downarrow & & \downarrow & & \\ 0 & \dashrightarrow & H & \dashrightarrow & E & \longrightarrow & G & \dashrightarrow & 0. \end{array}$$

The <u>exterior</u> <u>direct</u> <u>sum</u> [E] $\oplus$ [E'] of [E] $\in$ Ext(G,H) and [E'] $\in$

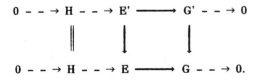

Ext(G',H') is given by [E ⊕ E'], where E ⊕ E' is the extension

$$0 \longrightarrow H \oplus H' \longrightarrow E \oplus E' \longrightarrow G \oplus G' \longrightarrow 0.$$

Finally, given [E], [E'] ∈ Ext(G,H), we define [E] + [E'] ∈ Ext(G,H) by

$$[E] + [E'] = \Delta^{*}\epsilon_{*}([E] \oplus [E'])$$

where $\Delta:G \longrightarrow G \oplus G$ is given by $\Delta(g) = (g,g)$, and $\epsilon:H\oplus H \longrightarrow H$ by $\epsilon(h_1,h_2) = h_1 + h_2$. Ext(G,H) is a group in which the identity element is [E], where E is any split extension

$$0 \longrightarrow H \rightleftarrows E \longrightarrow G \longrightarrow 0.$$

Given an element $[\alpha] \in \mathrm{Ext}(A)$, $\alpha:A \longrightarrow Q$ a proper injection, we obtain an exact sequence of C^{*}-algebras

$$(2.4) \qquad 0 \longrightarrow K \xrightarrow{j_\alpha} E \xrightarrow{\pi_\alpha} A \longrightarrow 0$$

as follows. We let $e_\alpha \in E$ be a projection with $\pi(e_\alpha) = \alpha(1)$, and

$$\mathcal{B}_\alpha = e_\alpha \mathcal{B} e_\alpha \cong \mathcal{B}(e_\alpha H) \cong \mathcal{B}$$

$$K_\alpha = e_\alpha K e_\alpha \cong K(e_\alpha H) \cong K.$$

Letting π'_α be the restriction of π to B_α, we let $E_\alpha = \pi'^{-1}_\alpha(\alpha A)$, $\pi_\alpha = \alpha^{-1} \cdot \pi'_\alpha$, and j_α be an isomorphism $K \cong K_\alpha$. We refer to (2.4) as an <u>extension</u> A by K. It is readily seen that if one choose another projection e'_α, one obtains an isomorphic extention.

From (2.4) we obtain an exact sequence

$$\begin{array}{ccccc} \mathbb{Z} \cong K_0(K) & \longrightarrow & K_0(E_\alpha) & \longrightarrow & K_0(A) \\ \uparrow i_\alpha & & & & \downarrow \\ K_1(A) & \longleftarrow & K_1(E_\alpha) & \longleftarrow & K_1(K) = 0 \end{array}$$

where

(2.5) $$i_\alpha([u]) = \mathrm{ind}_{\alpha(1)}\ \alpha(u) = \langle\alpha,u\rangle = \gamma_\infty([\alpha])([u]).$$

Thus it follows that if $[\alpha] \in \ker \gamma_\infty$, then we have a group extension E_α:

$$0 \longrightarrow \mathbb{Z} \longrightarrow K_0(E_\alpha) \longrightarrow K_0(A) \longrightarrow 0.$$

Lemma 2.5 The correspondence $\alpha \longrightarrow E_\alpha$ determines a natural homomorphism

$$\kappa:\ker \gamma_\infty \longrightarrow \mathrm{Ext}(K_0(A),\mathbb{Z}).$$

Proof If $\alpha \simeq \beta$, we have from (2.2) a unitary $u \in U_0(Q)$ such that $\beta = u\alpha u^*$. We may choose $\bar{u} \in U(\mathcal{B})$ with $\pi(\bar{u}) = u$ (use [T1, §4.8] and polar decomposition), and we may assume that e_β = ad $\bar{u}(e_\alpha)$ where

$$\mathrm{ad}\ \bar{u}(x) = \bar{u}x\bar{u}^*.$$

The map ad $\bar{u}$ provides an isomorphism of E_α onto E_β, and we have a commutative diagram

$$0 \longrightarrow K \begin{array}{c} \nearrow \\ \searrow \end{array} \begin{array}{c} E_\alpha \\ \Big\downarrow \mathrm{ad}\,\bar{u} \\ E_\beta \end{array} \begin{array}{c} \searrow \\ \nearrow \end{array} A \longrightarrow 0.$$

Assuming that $[\alpha] = [\beta] \in \ker \gamma_\infty$, it follows that we have a commutative diagram

$$0 \longrightarrow \mathbb{Z} \begin{array}{c} \nearrow \\ \searrow \end{array} \begin{array}{c} K_0(E_\alpha) \\ \Big\downarrow \\ K_0(E_\beta) \end{array} \begin{array}{c} \searrow \\ \nearrow \end{array} K_0(A) \longrightarrow 0$$

i.e., $[E_\alpha] = [E_\beta]$.

Given proper α, β with $[\alpha]$, $[\beta] \in \ker \gamma_\infty$ and $\alpha(1) \perp \beta(1)$, $\alpha(1) + \beta(1) < 1$, we let E be the C^*-algebraic pull-back given by the solid arrows in the resulting diagram

$$\begin{array}{ccccccccc} 0 & \dashrightarrow & K \oplus K & \overset{j}{\dashrightarrow} & E & \xrightarrow{p_2} & A & \dashrightarrow & 0 \\ & & \Big\| & & \Big\downarrow p_1 & & \Big\downarrow \Delta & & \\ 0 & \dashrightarrow & K \oplus K & \dashrightarrow & E_\alpha \oplus E_\beta & \longrightarrow & A \oplus A & \dashrightarrow & 0 \end{array}$$

where

$$E = \{((x,y),a) \in (E_\alpha \oplus E_\beta) \oplus A : \pi_\alpha(x) = \pi_\beta(y) = a\}$$

p_i are the projections, and $j(k,l) = (k,l,0)$ (see [B4]). In the corresponding K-theoretic diagram

$$\begin{array}{ccccccccc}
K_1(A) & \xrightarrow{j_*} & \mathbb{Z}\oplus\mathbb{Z} & \longrightarrow & K_0(E) & \longrightarrow & K_0(A) & \longrightarrow & 0 \\
\downarrow{\scriptstyle K_1(\Delta)} & & \| & & \downarrow & & \downarrow{\scriptstyle K_0(\Delta)} & & \\
K_1(A)\oplus K_1(A) & \xrightarrow[i_\alpha\oplus i_\beta]{} & \mathbb{Z}\oplus\mathbb{Z} & \longrightarrow & K_0(E_\alpha)\oplus K_0(E_\beta) & \longrightarrow & K_0(A)\oplus K_0(A) & \longrightarrow & 0
\end{array}$$

we have $K_i(\Delta) = \Delta$, and by assumption, $i_\alpha = i_\beta = 0$. It follows that $j_* = 0$, and it follows from [HS1, §III.1.2] that

$$0 \longrightarrow \mathbb{Z}\oplus\mathbb{Z} \longrightarrow K_0(E) \longrightarrow K_0(A) \longrightarrow 0$$

is just $\Delta^*(E_\alpha \oplus E_\beta)$.

We have a natural injection $E \longrightarrow E_{\alpha+\beta}$ given by $((x,y),a) \longrightarrow x + y$ and a commutative diagram

$$\begin{array}{ccccccccc}
0 & \longrightarrow & \mathcal{K}_\alpha\oplus\mathcal{K}_\beta & \longrightarrow & E & \longrightarrow & A & \longrightarrow & 0 \\
 & & \downarrow & & \downarrow & & \| & & \\
0 & \longrightarrow & \mathcal{K}_{\alpha+\beta} & \longrightarrow & E_{\alpha+\beta} & \longrightarrow & A & \longrightarrow & 0.
\end{array}$$

In the corresponding commutative diagram

$$\begin{array}{ccccccccc}
K_1(A) & \overset{j}{\dashrightarrow} & \mathbb{Z}\oplus\mathbb{Z} & \longrightarrow & K_0(E) & \dashrightarrow & K_0(A) & \dashrightarrow & 0 \\
\| & & \downarrow{\scriptstyle \epsilon} & & \downarrow & & \| & & \\
K_1(A) & \overset{i_{\alpha+\beta}}{\dashrightarrow} & \mathbb{Z} & \longrightarrow & K_0(E_{\alpha+\beta}) & \dashrightarrow & K_0(A) & \dashrightarrow & 0
\end{array}$$

we have $j = 0$, $\epsilon(x,y) = x + y$, and

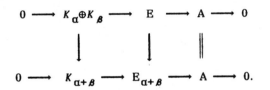

$$i_{\alpha+\beta} = \gamma_\infty([\alpha + \beta]) = \gamma_\infty([\alpha]) + \gamma_\infty([\beta]) = 0.$$

It follows from the dual of [HS1, §III.1.2] that $K_0(E_{\alpha+\beta})$ is the pushout given by the solid lines, i.e., the sequence

$$0 \longrightarrow \mathbb{Z} \longrightarrow K_0(E_{\alpha+\beta}) \longrightarrow K_0(A) \longrightarrow 0$$

represents

$$[E_\alpha] + [E_\beta] = \epsilon^* \Delta_*([E_\alpha] \oplus [E_\beta]).$$

Given a homomorphism $\varphi : A \longrightarrow B$ of purely infinite simple C^*-algebras A, B, we must check that

$$(2.6) \qquad \begin{array}{ccc} \ker \gamma_\infty^B & \xrightarrow{\ Ext(\varphi)\ } & \ker \gamma_\infty^A \\ \Big\downarrow \kappa & & \Big\downarrow \kappa \\ Ext(K_0(B),\mathbb{Z}) & \xrightarrow[\ K_0(\varphi)^*\ ]{} & Ext(K_0(A),\mathbb{Z}) \end{array}$$

where $h^*: Ext(H,\mathbb{Z}) \longrightarrow Ext(G,\mathbb{Z})$ is a functorially defined for any homomorphism $h: G \longrightarrow H$. We leave this straightforward calculation to the reader. ■

3. THE UNIVERSAL COEFFICIENT THEOREM FOR CUNTZ-KRIEGER ALGEBRAS.

Suppose that $\underline{a} = [a_{ij}]$ is a q x q matrix of 0's and 1's, no column or row of which contains only 0's. The Cuntz-Krieger algebra $0_{\underline{a}}$ may be characterized by the following properties: It is generated by a distinguished set of partial isometries s_i ($1 \leqslant i \leqslant q$) such that if $f_i = s_i{}^*s_i$ and $e_i = s_is_i{}^*$.

$$(3.1) \qquad 1 = \Sigma f_i, \quad e_i = \Sigma a_{ij}f_j.$$

In addition, it is generic with respect to these properties in the following sense. If A is any C^*-algebra with orthogonal projections f_i' and partial isometries $s_i':f_i' \simeq \Sigma a_{ij}f_j'$, there is a (unique) homomorphism φ: $O_{\underline{a}} \longrightarrow A$ with $\varphi(f_i) = f_i'$ and $\varphi(s_i) = s_i'$. It is evident that these properties uniquely determine the C^*-algebra $O_{\underline{a}}$. A general construction for such algebras $O_{\underline{a}}$ may be found in [E3], but for irreducible matrices this is unnecessary (see below).

We recall that $\underline{a}$ is said to be <u>irreducible</u> if for all i and j, there exists a $k > 0$ such $(\underline{a}^k)_{ij} \neq 0$. A matrix $\underline{a}$ is a <u>cyclic permutation</u> if after one permutes rows and columns it assumes the form

$$\pi = \begin{bmatrix} 0 \cdot\cdot\cdot\cdot\cdot\cdot\cdot\cdot\cdot\cdot 1 \\ 1\,0 \cdot\cdot\cdot\cdot\cdot\cdot\cdot\cdot\cdot\cdot \\ \cdot\cdot\cdot\cdot\cdot\cdot\cdot\cdot\cdot\cdot\cdot\cdot \\ \cdot\cdot\cdot\cdot\cdot\cdot\cdot\cdot 1\,0 \cdot\cdot \\ 0 \cdot\cdot\cdot\cdot\cdot\cdot\cdot\cdot\cdot 1\,0 \end{bmatrix}$$

If $\underline{a}$ is a non-cyclic irreducible matrix, $O_{\underline{a}}$ is simple [CK]. As a result any C^*-algebra A which is generated by partial isometries s_i satisfying (3.1) may be used as a model for $O_{\underline{a}}$. We may, for example , let $f_1,\dots,f_n$ be infinite dimensional projections in B with $\Sigma f_i = 1$. Then there exists $s_i:\Sigma a_{ij}f_j \simeq f_i$, and we may let $O_{\underline{a}}$ be the C^*-algebra they generate. On the other hand, if $\underline{a} = \pi$, we may let $O_{\underline{a}} = C(S^1) \otimes M_p$, with $f_i = 1 \otimes e_{ii}$, $s_i = 1 \otimes e_{i\ i+1}$ (see [E3]).

Throughout the remainder of this section, we shall assume that $\underline{a}$ is a <u>non-cyclic</u> <u>irreducible</u> <u>matrix</u>. From [C2] we have that $O_{\underline{a}}$ is simple and purely infinite, hence the theory of §2 is applicable.

Letting a^T be the transpose of a, the generators s_i determine isomorphisms, [C1].

$$K_0(O_{\underline{a}}) \cong Z^q/(1 - \underline{a}^T)\mathbb{Z}^q ,$$

(3.2)
$$K_1(O_{\underline{a}}) \cong \ker(1 - \underline{a}^T) ,$$

$$\mathrm{Ext}(O_{\underline{a}}) \cong \mathbb{Z}^q/(1 - \underline{a})Z^q ,$$

and under these isomorphisms the pairing

(3.3)
$$< , >:\mathrm{Ext}(O_{\underline{a}}) \times K_1(O_{\underline{a}}) \longrightarrow \mathbb{Z}$$

is just the pairing

$$< , >:(\mathbb{Z}^q/(1 - \underline{a})\mathbb{Z}^q) \times \ker(1 - \underline{a}^T) \longrightarrow \mathbb{Z}$$

induced from the dot product $\mathbb{Z}^q \times \mathbb{Z}^q \longrightarrow \mathbb{Z}$. In particular it follows that the corresponding map

$$\gamma_\infty:\mathrm{Ext}(O_{\underline{a}}) \longrightarrow \mathrm{Hom}(K_1(O_{\underline{a}}),\mathbb{Z})$$

is a surjection, since after a change of basis in Z^q we may assume that

$$1 - \underline{a} = 1 \oplus 1 \oplus \dots \oplus d_1 \oplus d_2 \oplus \dots \oplus 0 \dots \qquad (d_i \mid d_{i+1})$$

and thus

$$\mathrm{Ext}(O_{\underline{a}}) = \mathbb{Z}^q/(1 - \underline{a})\mathbb{Z}^q = 0 \oplus \dots \oplus \mathbb{Z}/d_1\mathbb{Z} \oplus \dots \oplus \mathbb{Z} \oplus \dots$$

$$K_1(O_{\underline{a}}) = \ker(1 - \underline{a}^T) = 0 \oplus \ldots \oplus 0 \oplus \ldots \oplus \mathbf{Z} \oplus \ldots$$

It is also clear from these formulas that

(3.4) $$\ker \gamma_\infty = T\mathrm{Ext}(O_{\underline{a}}) \qquad (T = \text{torsion subgroup}).$$

What is more important, however, is the following Universal Coefficient Theorem.

Theorem 3.1 (see [C1]) The natural map

$$\kappa : \ker \gamma_\infty \longrightarrow \mathrm{Ext}(K_0(O_{\underline{a}}), \mathbf{Z})$$

is an isomorphism, i.e., we have an exact sequence

$$0 \longrightarrow \mathrm{Ext}(K_0(O_{\underline{a}}), \mathbf{Z}) \longrightarrow \mathrm{Ext}(O_{\underline{a}}) \longrightarrow \mathrm{Hom}(K_1(O_{\underline{a}}), \mathbf{Z}) \longrightarrow 0$$

Proof We have for any finitely generated abelian group G that

(3.5) $$\mathrm{Ext}(G, \mathbf{Z}) = (TG)^*$$

where $(TG)^*$ is the Pontryagin dual for the finite group TG. Since $\ker \gamma_\infty = T\mathrm{Ext}(O_{\underline{a}})$ is a finite group of the same order as $TK_0(O_{\underline{a}})^*$, it suffices to prove that κ is one-one.

Let us suppose that $[\alpha] \in \mathrm{Ext}(O_{\underline{a}})$ is such that $\gamma_\infty[\alpha] = 0$, and $\kappa([\alpha]) = 0$, i.e., we have a split exact sequence

$$0 \longrightarrow \mathbf{Z} \longrightarrow K_0(E_\alpha) \underset{\longrightarrow}{\overset{\rho}{\longleftarrow}} K_0(O_{\underline{a}}) \longrightarrow 0.$$

Note that $K_0(O_{\underline{a}})$ is generated by $[f_1], \ldots, [f_q]$, and it is evident from Lemma 2.2 that we may choose orthogonal projections $f_j' \in E_\alpha$ with

$\rho([f_j]) = [f_j']$. We have

$$[\Sigma a_{ij} f_j'] = \Sigma a_{ij}[f_j'] = [f_i']$$

since the latter equality holds for the $[f_j]$, hence we may select partial isometries $v_i : \Sigma a_{ij} f_j' \simeq f_i'$. We have that

$$u_i = \pi(v_i)\alpha(s_i^*) \in U(f_i O_{\underline{a}} f_i)$$

satisfies $\pi(v_i) = u_i \alpha(s_i)$. Also since $\gamma_\infty[\alpha] = 0$, we have from (2.5) that $\mathrm{ind}[u_i + (1 - \alpha(f_i))] = 0$. Regarding u_i as an element of $Q(f_i' H)$. we have from Lemma 2.1 that u_i has index 0, and thus there is a $u_i' \in U(B(f_i H))$ with $\pi_\alpha(u_i') = u_i$. We may regard u_i' as an element of $U(f_i' E_\alpha f_i')$, and the partial isometries $s_i' = u_i'^* v_i$ satisfy $s_i' : \Sigma a_{ij} f_j' \simeq f_i'$ and $\pi(s_i') = \alpha(s_i)$. Defining $\tau : O_{\underline{a}} \longrightarrow E_\alpha$ by $\tau(s_i) = s_i'$, we have that τ is a homomorphic lifting of α, and thus $[\alpha] = 0$. ■

From the above discussion we have an isomorphism

$$\kappa : \mathrm{TExt}(O_{\underline{a}}) \cong (\mathrm{TK}_0(O_{\underline{a}}))^* \tag{3.6}$$

where * indicates the Pontryagin dual. This is <u>natural</u> in the sense that for any $\varphi : O_{\underline{a}} \longrightarrow O_{\underline{b}}$, we have from (2.6) that

$$\begin{array}{ccc} \mathrm{TExt}(O_{\underline{b}}) & \xrightarrow{\mathrm{TExt}(\varphi)} & \mathrm{TExt}(O_{\underline{a}}) \\ \downarrow \kappa & & \downarrow \kappa \\ \mathrm{TK}_0(O_{\underline{b}})^* & \xrightarrow{\mathrm{TK}_0(\varphi)^*} & \mathrm{TK}_0(O_{\underline{a}})^* \end{array} \tag{3.7}$$

is commutative. For the bottom row we note that (3.3) is functorial.

4. THE CUNTZ FORMULA.

Throughout this section we let $A = O_{\underline{a}}$ and $B = O_{\underline{b}}$, where $\underline{a}$ and $\underline{b}$ are non-cyclic, irreducible qxq and rxr matrices. We let f_i, s_i $(1 \leqslant i \leqslant q)$ denote the usual generators of A.

Given a proper homomorphism $\varphi:A \longrightarrow B$, we have a corresponding homomorphism $K_0(\varphi):K_0(A) \longrightarrow K_0(B)$. Since $\varphi \simeq \psi$ implies that $K_0(\varphi) = K_0(\psi)$, and $\varphi \perp \psi$, $\varphi(1) + \psi(1) < 1$ imply

$$K_0(\varphi + \psi) = K_0(\varphi) + K_0(\psi),$$

we obtain an additive map

$$K_0:[A,B]_p \longrightarrow \mathrm{Hom}(K_0(A),K_0(B)). \tag{4.1}$$

To see that this is a surjection, suppose that $\theta:K_0(A) \longrightarrow K_0(B)$ is given. We may let $\varphi([f_i]) = [\bar{f}_i]$ where $\bar{f}_i \in \mathrm{Proj}\ B$ are orthogonal. Since $[f_i] = [\Sigma a_{ij}f_j]$, it follows that $[\bar{f}_i] = [\Sigma a_{ij}\bar{f}_j]$ and thus we may find partial isometries $\bar{s}_i:\Sigma a_{ij}\bar{f}_j \simeq \bar{f}_i$. We may thus define $\varphi:A \longrightarrow B$ by letting $\varphi(s_i) = \bar{s}_i$. It is immediate that $K_0(\varphi) = \theta$.

Our next task is to determine the equivalence relation on $[A,B]_p$ determined by K_0.

<u>Lemma</u> <u>4.1</u> Suppose that $\varphi,\psi:A \longrightarrow B$ are proper injections such that $K_0(\varphi) = K_0(\psi)$. Then there is a proper $\psi':A \longrightarrow B$ such that $\psi \simeq \psi'$ and $\psi'(f_i) = \varphi(f_i)$ $(1 \leqslant i \leqslant q)$.

Proof Since $[\varphi(f_i)] = K_0(\varphi)[f_i] = [\psi(f_i)]$, we may choose partial isometries $v_i:\psi(f_i) \simeq \varphi(f_i)$. Then $v = \Sigma v_i:\psi(1) \simeq \varphi(1)$ and from Corollary 2.4 there exists a unitary $u \simeq 1$ in B such that $u\psi(f_i)u^* = \varphi(f_i)$. It follows that $\psi' = u\psi u^*$ has the desired properties. ■

Given proper $\varphi,\psi:A \longrightarrow B$ with $\varphi(f_i) = \psi(f_i)$ $(1 \leqslant i \leqslant q)$ we let $\bar{f}_i = \varphi(f_i)$. Then $v_i = \psi(s_i)\varphi(s^*_i)$ is a unitary in $\bar{f}_iB\bar{f}_i$. Conversely, given $v_i \in U(\bar{f}_iB\bar{f}_i)$, we obtain a proper homomorphism $\psi:A \longrightarrow B$ by letting $\psi(s_i) = v_i\varphi(s_i)$. If $v_i \simeq \bar{v}_i$ in $\bar{f}_iB\bar{f}_i$ and $\bar{\psi}$ is the corresponding map it readily follows that $\psi \simeq \bar{\psi}$.

We define an action of $K_1(B)^q$ on $[A,B]_p$ as follows. Given $\mathbf{d} = ([u_1],...,[u_q]) \in K_1(B)^q$ and $[\varphi] \in [A,B]_p$, we let $\bar{f}_i = \varphi(f_i)$. From Lemma 2.1 we may choose $v_i \in U(\bar{f}_iB\bar{f}_i)$ with $u_i \simeq v_i + (1 - \bar{f}_i)$. We define $\psi:A \longrightarrow B$ by $\psi(s_i) = v_i\varphi(s_i)$. From Lemma 2.1, if $u_i \simeq \bar{v}_i + (1 - \bar{f}_i)$, then $\bar{v}_i \simeq v_i$ in $U(\bar{f}_iB\bar{f}_i)$, and thus if $\bar{\psi}(s_i) = \bar{v}_i\varphi(s_i)$, $[\bar{\psi}] = [\psi]$. Given a path of proper homomorphisms φ^t $(0 \leqslant t \leqslant 1)$ joining $\varphi = \varphi^0$ to φ^1, we let $\bar{f}_i^t = \varphi^t(f_i)$. From [EK1, Lemma 3.9] applied to $A = \Sigma\mathbb{C}\bar{f}_i + \mathbb{C}(1-\Sigma\bar{f}_i)$ there is a path $w^t \in U(B)$ with $w^t\bar{f}_i(w^t)^* = \bar{f}_i^t, w^0 = 1$. We have that $v_i^t = w^tv_iw^{t*} \in U(\bar{f}_i^tB\bar{f}_i^t)$, and we may thus define $\psi^t:A \longrightarrow B$ by $\psi^t(s_i) = v_i^t\varphi^t(s_i)$. This is a path, hence $\psi \simeq \psi^1$. Since

$$u_i \simeq w^1u_iw^{1*} \simeq v_i^1 + (1 - \bar{f}_i^1),$$

$[\psi^1]$ is the class determined by $\mathbf{d} = ([u_1],...,[u_q])$ and φ^1. Thus, $[\psi]$ only depends on $\mathbf{d}$ and $[\varphi]$, and we write $[\psi] = \mathbf{d}[\varphi]$.

Lemma 4.2 [C1] Given $\mathbf{d} = (d_1,\ldots,d_p) \in K_1(B)^q$ and $[\varphi] \in [A,B]$, we have $\mathbf{d}[\varphi] = [\varphi]$ if and only if $\mathbf{d} \in (1 - \mathbf{a})\, K_1(B)^q$. i.e.. there exists $\mathbf{k} = (k_1,\ldots,k_q) \in K_1(B)^q$ such that for each i,

(4.2) $$d_i = k_i - \Sigma a_{ij}k_j \qquad (1 \leqslant i \leqslant q)$$

Proof Let us suppose that d_i and k_i satisfy (4.2) and $\varphi{:}A \longrightarrow B$ is arbitrary. Let $d_i = [u_i]$, $u_i \in U(B)$ and letting $\bar{f}_i = \varphi(f_i)$, we choose $v_i \in U(\bar{f}_i B\bar{f}_i)$ with $u_i \simeq v_i + (1 - \bar{f}_i)$. By definition $\mathbf{d}[\varphi] = [\psi]$ where $\psi(s_i) = v_i\varphi(s_i)$, and we must prove that $\psi \simeq \varphi$.

From Lemma 2.1 we may choose $w_i \in U(\bar{f}_i B\bar{f}_i)$, $(1 \leqslant i \leqslant q)$ such that if $w_i' = w_i + (1 - \bar{f}_i)$, $[w_i'] = -k_i$. Letting $\bar{f} = 1 - \Sigma f_i$ we may choose $w_0 \in U((1 - \bar{f})B(1 - \bar{f}))$ such that $[w_0'] = \Sigma k_i$ and thus $w = w_0'\ldots w_q'$ is in $U_0(B)$. It follows that $\psi' = w\psi w^* \simeq \psi$. and it suffices to show that $\psi' \simeq \varphi$. Noting that $\psi'(f_i) = \psi(f_i) = \varphi(f_i) = \bar{f}_i$. this will follow if we show that each unitary $u_i = \psi'(s_i)\varphi(s_i^*)$ is equivalent to the identity in $U(\bar{f}_i B\bar{f}_i)$.

Given a proper projection $e \in B$ and $v \in U(eBe)$, we let $[v]_e$ be the class of $v' = v + (1 - e)$ in $U(B)$. If $v_i \in U(e_iBe_i)$ $(i = 1,2)$ where $e_1 \perp e_2$ and $e = e_1 + e_2 < 1$, then $v_1 + v_2 \in U(eBe)$ and $v' = v_1'v_2'$. It follows that

(4.3) $$[v]_e = [v_1]_{e_1} + [v_2]_{e_2}.$$

Also, if $s{:}e \simeq f$ where $e,f \in \mathrm{Proj}_p B$, we may choose $w \in U(B)$ with $we = s$ and $wew^* = f$. If $v \in U(eBe)$, then $(wvw^*)' = wv'w^*$, and thus

$$(4.4) \qquad [svs^*]_f = [wv'w^*] = [v'] = [v]_e.$$

In the following computation we use the fact that $w\bar{f}_i = w_i$ and that both $\varphi(s_i)$ and $\psi(s_i)$ are partial isometries with domain $\Sigma a_{ij}\bar{f}_j$ and range $\bar{f}_i$. We have

$$\begin{aligned}
(4.5) \quad [\psi'(s_i)\varphi(s_i^*)]_{\bar{f}_i} &= [w_i\psi(s_i)w^*\varphi(s_i^*)]_{\bar{f}_i} \\
&= [w_i\psi(s_i)(\Sigma a_{ij}w_j^*)\varphi(s_i^*)]_{\bar{f}_i} \\
&= [w_i\psi(s_i)\varphi(s_i^*)\varphi(s_i)(\Sigma a_{ij}w_j^*)\varphi(s_i^*)]_{\bar{f}_i} \\
&= [w_i]_{\bar{f}_i} + [\psi(s_i)\varphi(s_i^*)]_{\bar{f}_i} + [\varphi(s_i)(\Sigma a_{ij}w_j^*)\varphi(s_i^*)]_{\bar{f}_i} \\
&= -k_i + d_i + [\Sigma a_{ij}w_j^*]_{\Sigma a_{ij}\bar{f}_j} \\
&= -k_i + d_i + \Sigma a_{ij}k_j \\
&= 0
\end{aligned}$$

Thus we indeed have u_i is equivalent to the identity in $\bar{f}_iB\bar{f}_i$.

Conversely, suppose that $d[\varphi] = [\varphi]$. To show that $d \in (1-\underline{a})K_1(B)^q$, it suffices to prove that for all $r \in Ext(B)$,

$$(4.6) \qquad (\langle r,d_1\rangle,\dots,\langle r,d_q\rangle) \in (1 - \underline{a})\mathbb{Z}^q$$

To see this we note that (3.2) implies that $K_1(B) = \mathbb{Z}^s$ $(s \leqslant r)$ and for each h $(1 \leqslant h \leqslant s)$ there is an $r^h \in Ext(B)$ such that if $d = (d_1,\dots,d_s) \in K_1(B)$, $r^h(d) = d_h$. Letting $r = r^h$ in (4.6), it follows that for each h, there exists $k_1{}^h,\dots,k_q{}^h \in \mathbb{Z}$ such that

$$r^h(d_i) = k_i{}^h - \Sigma\, a_{ij}k_j{}^h.$$

Letting $k_i = (k_i^1,\dots,k_i^s) \in \mathbf{Z}^s$, we obtain (4.2).

We let $d_i = [v_i]_{\bar{f}_i}$, $v_i \in U(\bar{f}_i B \bar{f}_i)$. By assumption the map defined by $\psi(s_i) = v_i\, \varphi(s_i)$ satisfies $\psi \simeq \varphi$. It follows that if $r = [\beta]$, $\beta: B \longrightarrow Q$ a proper homomorphism, then

$$\beta \circ \psi \simeq \beta \circ \varphi : B \longrightarrow Q.$$

Since $\beta\circ\psi(s_i) = \beta(v_i)\beta\circ\varphi(s_i)$, it follows from the calculation of $\mathrm{Ext}(O_{\underline{b}})$ in [CK1] that

$$\langle r, d_i\rangle = \mathrm{ind}_{\beta(1)}\ \beta(v_i) \in (1 - \underline{a})\mathbf{Z}^q. \quad \blacksquare$$

Corollary 4.3 (see [C1]) Given proper homomorphisms φ,ψ: $A \longrightarrow B$ we have $\varphi \simeq \psi$ if and only if $K_0(\varphi) = K_0(\psi)$ and $\mathrm{Ext}(\varphi) = \mathrm{Ext}(\psi)$.

Proof Suppose that $K_0(\varphi) = K_0(\psi)$ and

$$\varphi^* = \mathrm{Ext}(\varphi) = \mathrm{Ext}(\psi) = \psi^*.$$

From Lemma 3.1 we may assume that $\psi(f_i) = \varphi(f_i) = \bar{f}_i$, and let

$$v_i = \psi(s_i)\varphi(s_i^*) \in U(\bar{f}_i B \bar{f}_i).$$

Given $\beta \in \mathrm{Ext}(B)$ we have $[\beta\circ\psi] = \psi^*[\beta] = \varphi^*[\beta] = [\beta\circ\varphi]$, and thus, from [C1], $\beta(v_i) \in (1 - \underline{a})\mathbf{Z}^q$. As above it follows that $\mathbf{d} = ([v_1'],\dots,[v_q']) \in (1 - \underline{a})K_1(B)^q$, and thus from Lemma 4.2, $[\psi] = \mathbf{d}[\varphi] = [\varphi]$. The converse is trivial. $\blacksquare$

Our next task is to prove that $[A,B]_p$ is a group. We let $O_2 =$

$O_{\underline{a}}$ where $\underline{a} = \begin{bmatrix} 1 & 1 \\ 1 & 1 \end{bmatrix}$.

Lemma 4.4 $[O_2,O_2]_p$ consists of single element.

Proof From (3.1), $K_0(O_2) = \mathrm{Ext}(O_2) = 0$. Thus for any proper $\varphi,\psi:A \longrightarrow B$, $K_0(\varphi) = K_0(\psi)$ and $\mathrm{Ext}(\varphi) = \mathrm{Ext}(\psi)$. Thus from Corollary 4.3, $\varphi \simeq \psi$.

Lemma 4.5 There is a proper map $\tau:A \longrightarrow B$ such that $[\tau] + [\tau] = [\tau]$.

Proof Letting $\tau_0:O_2 \longrightarrow O_2$ be a proper injection (which exists since (4.1) is surjective), we have from Lemma 4.4 that $[\tau_0] + [\tau_0] = [\tau_0]$. Again since (4.1) is surjective, there exist proper homomorphisms $\varphi:A \longrightarrow O_2$ and $\psi:O_2 \longrightarrow B$. Then $\tau = \psi\tau_0\varphi$ is the desired map. ∎

Lemma 4.6 $[A,B]_p$ is a group with identity $[\tau]$.

Proof Since $K_0(\tau) + K_0(\tau) = K_0(\tau)$ and $\mathrm{Hom}(K_0(A),K_0(B))$ is a group, $K_0(\tau) = 0$. Similarly $\mathrm{Ext}(\tau) = 0$. Given a proper homomorphism $\varphi:A \longrightarrow B$ with $\varphi(1) \perp \tau(1)$, $\varphi(1) + \tau(1) < 1$, it follows that $K_0(\varphi+\tau) = K_0(\varphi)$ and $\mathrm{Ext}(\varphi+\tau) = \mathrm{Ext}(\varphi)$, and thus $[\varphi] + [\tau] = [\varphi]$.

Given a proper homomorphism $\varphi:A \longrightarrow B$, we may use the surjectivity of K_0 in (4.1) to find a proper homomorphism ψ with $K_0(\psi) = -K_0(\varphi)$, $\psi(1) \perp \varphi(1)$, $\psi(1) + \varphi(1) < 1$. It follows that $K_0(\varphi + \psi) = K_0(\tau)$ and from Lemma 3.1, we may assume that $\tau(f_i) = (\varphi + \psi)(f_i) = \bar{f}_i$. We have

$$v_i = (\varphi + \psi)(s_i)\tau(s_i^*) \in U(\bar{f}_i B \bar{f}_i).$$

We choose a proper τ' with $\tau' \simeq \tau$, $\tau(1) \perp \tau'(1)$, $\tau(1) + \tau'(1) < 1$, and we let $\tau'(f_i) = \bar{f}_i'$. Since $[\bar{f}_i] = [\tau(f_i)] = 0$. and similarly $[\bar{f}_i'] = 0$, we have $\bar{f}_i' \simeq f_i$. Thus letting $e_i = (\bar{f}_i + \bar{f}_i')$, $e_i B e_i \cong M_2(\bar{f}_i B \bar{f}_i)$, and we let $w_i \in e_i B e_i$ correspond to $\begin{bmatrix} 0 & 0 \\ 0 & v_i^* \end{bmatrix}$. It follows that $v_i + w_i$ is homotopic to the identity in $e_i B e_i$. Defining $\theta : A \longrightarrow B$ by $\theta(s_i) = w_i \tau'(s_i)$ we have

$$(\varphi + \psi + \theta)(s_i) = (v_i + w_i)(\tau + \tau')(s_i)$$

and thus $\varphi + \psi + \theta \simeq \tau + \tau'$, i.e., $[\varphi] + [\psi + \theta] = 0$. ■

We define $\lambda : K_1(B)^q \longrightarrow [A,B]_p$ by $\lambda(\mathbf{d}) = \mathbf{d}[\tau]$, where $[\tau] = 0$.

Lemma 4.7 λ is a homomorphism.

Proof We define $\tau' \simeq \tau$. $\bar{f}_i$, $\bar{f}_i'$, e_i $(i = 1,2)$ as in the proof of Lemma 4.6, and again identify $e_i B e_i$ with $M_2(\bar{f}_i B \bar{f}_i)$.

Given $\mathbf{c} = ([u_1],\ldots,[u_q])$ and $\mathbf{d} = ([v_1],\ldots,[v_q])$ in $K_1(B)^q$ we let $u_i \simeq x_i + (1 - \bar{f}_i)$, $v_i \simeq y_i + (1 - \bar{f}_i)$, $x_i, y_i \in U(f_i B f_i)$. Letting $y_i' \in U(\bar{f}_i' B \bar{f}_i')$ correspond to $\begin{bmatrix} 0 & 0 \\ 0 & y_i \end{bmatrix}$ in $M_2(\bar{f}_i B \bar{f}_i)$ we also have that

$$y_i + \bar{f}_i' \simeq y_i' + \bar{f}_i$$

in $U(e_i B e_i)$. It follows that in $U(B)$

$$v_i \simeq y_i' + (1 - \bar{f}_i'),$$

and

$$u_i v_i \simeq (x_i + (1 - \bar{f}_i))(y_i' + (1 - \bar{f}_i')) + (1 - e_i)$$

$$= (x_i + y_i') + (1 - e_i).$$

By definition, $\lambda(c) = [\varphi]$, $\lambda(d) = [\psi]$, and $\lambda(cd) = [\theta]$ where

$$\varphi(s_i) = x_i \tau(s_i)$$

$$\psi(s_i) = y_i' \tau'(s_i)$$

$$\theta(s_i) = (x_i + y_i')(\tau + \tau')(s_i) = \varphi(s_i) + \psi(s_i),$$

since $[\tau] = [\tau'] = [\tau + \tau']$. It follows that

$$\lambda(cd) = [\theta] = [\varphi + \psi] = \lambda(c) + \lambda(d).$$ ■

Tensoring the exact sequence

$$0 \longrightarrow (1 - \underline{a})\mathbf{Z}^q \longrightarrow \mathbf{Z}^q \longrightarrow \mathrm{Ext}(A) \longrightarrow 0$$

with the free abelian group $K_1(B)$, we obtain the exact sequence

$$0 \longrightarrow (1 - \underline{a})K_1(B)^q \longrightarrow K_1(B)^q \longrightarrow \mathrm{Ext}(A) \otimes K_1(B) \longrightarrow 0$$

It follows that

$$\ker K_0 = K_1(B)^q/(1 - \underline{a})K_1(B)^q \cong \mathrm{Ext}(A) \otimes K_1(B)$$

and we have

Theorem 4.8 [C1] There is an exact sequence

$$(4.7)\quad 0 \longrightarrow K_1(B)\otimes \mathrm{Ext}(A) \longrightarrow [A,B]_p \xrightarrow{K_0} \mathrm{Hom}(K_0(A),K_0(B)) \longrightarrow 0$$

For a more explicit description of $[A,B]_p$, we consider the homomorphism

$$K_0\oplus \mathrm{Ext}:[A,B]_p \longrightarrow \mathrm{Hom}(K_0(A),K_0(B)) \oplus \mathrm{Hom}(\mathrm{Ext}(B),\mathrm{Ext}(A))$$

From Corollary 4.3 this is an injection. Using the identification (3.4) we obtain

Theorem 4.9 [C1] The image of $K_0\oplus\mathrm{Ext}$ consists of the pairs $g \oplus h$ such that

$$(4.8)\qquad (Tg)^* = Th.$$

Proof From (3.7) we have that if $\varphi:A \longrightarrow B$ is proper, then $g = K_0(\varphi)$ and $h = \mathrm{Ext}(\varphi)$ satisfy the given equation.

Conversely, suppose first that $g:K_0(A) \longrightarrow K_0(B)$ and $h:\mathrm{Ext}(B) \longrightarrow \mathrm{Ext}(A)$ satisfy $g = 0$ and $Th = 0$. Then on purely algebraic grounds we must have elements $d_i \in K_1(B)$ and $r_i \in \mathrm{Ext}(A)$ $(1 \leqslant i \leqslant q)$ with

$$h(t) = \Sigma\langle t,d_i\rangle r_i \qquad (t \in \mathrm{Ext}(B))$$

i.e., with the usual interpretation

$$h = \Sigma r_i \otimes d_i \in \mathrm{Ext}(A) \otimes K_1(B)$$

Letting $[\tau] \in [A,B]_p$ be the identity, and $\mathbf{d} = (d_1,\dots,d_q)$, we have

$$K_0(\mathbf{d}[\tau]) = K_0([\tau]) = 0.$$

Thus it suffices to prove that $\mathrm{Ext}(\mathbf{d}[\tau]) = h$.

We let $\tau(f_i) = \bar{f}_i$, $d_i = [u_i]$, and $u_i = v_i + (1 - \bar{f}_i)$, $v_i \in U(\bar{f}_i B \bar{f}_i)$. Then $\mathbf{d}[\tau] = [\psi]$, where $\psi(s_i) = v_i \tau(s_i)$. Letting $\delta = [\beta] \in \mathrm{Ext}(B)$, $\beta: B \longrightarrow Q$ proper, we have that

$$\mathrm{Ext}(\mathbf{d}[\tau])[\beta] = [\mathrm{Ext}(\psi)(\beta)] = [\beta \circ \psi].$$

The isomorphism $\mathrm{Ext}(A) \cong \mathbb{Z}^q/(1 - \underline{a})\mathbb{Z}^q$ (see (3.1)) is given by

$$[\alpha] \longrightarrow (\mathrm{ind}_{\alpha(f_i)} \alpha(s_i)\alpha_0(s_i^*)) \quad (\mathrm{mod}(1 - \underline{a})\mathbb{Z}^q),$$

where $\alpha_0: A \longrightarrow Q$ is any trivial proper map with $\alpha_0(f_i) = \alpha(f_i)$. Letting $r_i \in \mathrm{Ext}(A)$ correspond to the image of the basis vectors in $\mathbb{Z}^q$, this may be rewritten

$$[\alpha] = \Sigma \mathrm{ind}_{\alpha(f_i)} \alpha(s_i)\alpha_0(s_i^*) r_i.$$

Since $[\beta \circ \tau] = \tau^*[\beta] = 0$, we may let $\alpha = \beta \circ \psi$ and $\alpha_0 = \beta \circ \tau$. Then we have

$$\mathrm{ind}_{\beta(\bar{f}_i)} \beta \circ \psi(s_i)\ \beta \circ \tau(s_i^*) = \mathrm{ind}_{\beta(\bar{f}_i)} \beta(v_i) = \langle \beta, v_i \rangle = \langle \delta, d_i \rangle$$

and

$$\mathrm{Ext}(\mathbf{d}[\tau])(\delta) = \Sigma \langle \delta, d_i \rangle r_i = h(\beta).$$

Given general homomorphisms $g: K_0(A) \longrightarrow K_0(B)$ and $h: \mathrm{Ext}(B) \longrightarrow \mathrm{Ext}(A)$ satisfying (4.8), we may use the surjectivity of (4.1) to find a proper $\varphi: A \longrightarrow B$ such that $K_0(\varphi) = g$. Since $K_0(\varphi)$, $\mathrm{Ext}(\varphi)$ satisfy (4.8), the same is true for the pair $g_0 = g - K_0(\varphi) = 0$ and $h_0 = h - \mathrm{Ext}(\varphi)$, i.e., $Th_0 = 0$. Thus we may choose $\psi: A \longrightarrow B$ such that $K_0(\psi) = 0$, $\mathrm{Ext}(\psi) = h_0$. It is no restriction to assume that $\psi(1) \perp \varphi(1)$, and we have $(K_0 \oplus \mathrm{Ext})(\varphi + \psi) = (g,h)$. ■

Following Cuntz [C1], let us see why

$$K_0 \oplus K_1 : [A,B]_p \longrightarrow \mathrm{Hom}(K_0(A), K_0(B)) \oplus \mathrm{Hom}(K_1(A), K_1(B))$$

need not be faithful. Letting

$$\mathrm{FExt}(A) = \mathrm{Ext}(A)/\mathrm{TExt}(A)$$

(this may be regarded as a functor), it is apparent that $\langle\ ,\ \rangle$ induces a faithful pairing

$$\langle\ ,\ \rangle : \mathrm{FExt}(A) \times K_1(A) \longrightarrow \mathbb{Z}.$$

Letting $\mathrm{Hom}(F,\mathbb{Z}) = F^d$ for any finitely generated abelian group (this can also be regarded as a functor), we have a natural isomorphism

(4.9) $$\mathrm{FExt}(A) \cong K_1(A)^d.$$

Given finitely generated abelian groups G, H, fix splittings $G \cong FG \oplus TG$, $H \cong FH \oplus TH$. Then any homomorphism $h: G \longrightarrow H$ has a matrix representation

$$h = \begin{bmatrix} Fh & 0 \\ h_{21} & Th \end{bmatrix}$$

where $h_{21}: FG \longrightarrow TH$ (note that O is the only homomorphism $TG \longrightarrow FH$).

Given a proper homomorphism $\varphi:A \longrightarrow B$, we may apply this to the homomorphism

$$\mathrm{Ext}(\varphi):\mathrm{Ext}(B) \longrightarrow \mathrm{Ext}(A).$$

Owing to the fact that (3.6) and (4.9) are natural, we have

$$\mathrm{Ext}(\varphi) = \begin{bmatrix} K_1(\varphi)^d & 0 \\ h_{21} & (TK_0(\varphi))^* \end{bmatrix}$$

From Theorem 4.9, given arbitrary homomorphisms $h_i:K_i(A) \longrightarrow K_i(B)$ ($i = 0,1$), and an arbitrary h_{21}, there exists a φ with $h_i = K_i(\varphi)$ and $\mathrm{Ext}(\varphi) = \begin{bmatrix} h_1^d & 0 \\ h_{21} & (Th_0)^* \end{bmatrix}$. In particular, if $h_1 = h_2 = 0$, and $h_{21} \neq 0$, $[\varphi] \neq 0$ but $K_0 \oplus K_1(\varphi) = 0$. It is a simple matter to produce a corresponding Cuntz-Krieger example.

5. SYSTEMS OF GROUPS.

Given a category $\underline{C}$, we define the category dir $\underline{C}$ of direct sequences in $\underline{C}$ as follows. An object (C_n,φ_n) in $\underline{C}$ is a diagram in $\underline{C}$ of the form

$$C_1 \xrightarrow{\varphi_1} C_2 \xrightarrow{\varphi_2} \cdots \tag{5.1}$$

A <u>morphism</u> $(f_n,p(n)):(C_n,\varphi_n) \longrightarrow (D_n, \psi_n)$ consists of a sequence of maps $f_n:C_n \longrightarrow D_{p(n)}$, $p(1) < p(2) < \ldots$, such that the following diagram commutes:

$$\begin{array}{ccccccc} C_1 & \longrightarrow & C_2 & \longrightarrow & \cdots & & \\ \downarrow f_1 & & f_2 & & & & \\ D_{p(1)} & \longrightarrow & \cdots & \longrightarrow & D_{p(2)} & \longrightarrow & \cdots \end{array}$$

We use the notation

$$\varphi_{mn} = \varphi_{m-1} \circ \ldots \circ \varphi_n : C_n \longrightarrow C_m \ (n < m), \ \varphi_{nn} = \text{id},$$

and we will also employ the abbreviations (C_n), (f_n).

Two morphisms $(f_n, p(n))$, $(g_n, q(n)) : (C_n, \varphi_n) \longrightarrow (D_n, \psi_n)$ are said to be <u>equivalent</u>, $(f_n, p(n)) \simeq (g_n, q(n))$, if for each n, there is an $r \geqslant p(n), q(n)$ with

$$\psi_{rp(n)} \circ f_n = \psi_{rq(n)} \circ g_n.$$

Finally, the <u>injective category</u> inj <u>C</u> consists of the objects of dir <u>C</u> with the morphisms $(C_n, \varphi_n) \longrightarrow (D_n, \psi_n)$ being the equivalence classes of such morphisms in dir <u>C</u>. This definition is quite analogous to the homotopy category of topological spaces, in which one uses the same objects but considers homotopy classes of maps.

One defines the <u>pro-category</u> pro <u>C</u> in a dual fashion, starting with the inverse category inv <u>C</u> of diagrams

(5.2) $$\cdots \xrightarrow{\varphi_2} C_2 \xrightarrow{\varphi_1} C_1,$$

denoted (C_n, φ_n), and morphisms $(f_n, p(n))$ a sequence of homomorphism $f_n : C_{p(n)} \longrightarrow D_n$ for which one has a commutative diagram

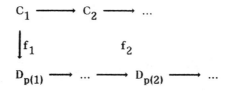

$$\begin{array}{ccccccc} \cdots \longrightarrow & C_{p(2)} & \longrightarrow & \cdots & \longrightarrow & C_{p(1)} \\ & & & f_2 & & \downarrow f_1 \\ & & \cdots \longrightarrow & D_2 & \longrightarrow & D_1 \end{array}$$

We say $(f_n,p(n))$ is <u>equivalent</u> to $(g_n,q(n))$, $(f_n,p(n)) \simeq (g_n,q(n))$, if for each n there is an $r \geqslant p(n),q(n)$ with

$$f_n \circ \varphi_{p(n)r} = g_n \circ \varphi_{q(n)r}$$

where

$$\varphi_{mn} = \varphi_m \circ \ldots \circ \varphi_{n-1}: C_n \longrightarrow C_m \ (n > m), \ \varphi_{nn} = \text{id}.$$

The objects in pro $\underline{C}$ are then just those in inv $\underline{C}$, and the morphisms are equivalence classes of morphisms in inv $\underline{C}$.

Given a direct system of abelian groups

(5.3) $$G_1 \xrightarrow{\varphi_1} G_2 \xrightarrow{\varphi_2} \cdots$$

there is a <u>direct</u> <u>limit</u> $(\varinjlim G_n, \varphi_\infty)$. This consists of a fixed group $G_\infty = \varinjlim G_n$ together with homomorphisms $\varphi_{n\infty}: G_n \longrightarrow G_\infty$ with the usual universal property. Furthermore, any morphism $(f_n,p(n)):(G_n,\varphi_n) \longrightarrow (H_n,\psi_n)$ induces a homomorphism.

$$\varinjlim(f_n,p(n)): \varinjlim G_n \longrightarrow \varinjlim H_n$$

and if $(f_n,p(n)) \simeq (g_n,q(n))$, then $\varinjlim(f_n,p(n)) \cong \varinjlim(g_n,q(n))$. Thus letting $\underline{A}$ be the category of abelian groups, we obtain a functor

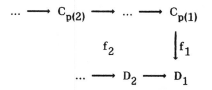

$$\underrightarrow{\lim}\text{:inj } \underline{A} \longrightarrow \underline{A}$$

The "shape question" makes sense in this algebraic setting. Thus we may ask whether or not a homomorphism of limit groups is induced by a homomorphism of inverse system. For sequences of finitely generated groups (5.3) this is the case. In fact we have

__Lemma__ __5.1__ Given two direct systems of finitely generated abelian groups (G_n,φ_n), (H_n, ψ_n), and a homomorphism $h:G \longrightarrow H$, there exists a homomorphism $(f_n,p(n)):(G_n,\varphi_n) \longrightarrow (H_n,\psi_n)$ such that $h = \underrightarrow{\lim}(f_n,p(n))$. If $(g_n,q(n))$ is another such homomorphism, then $(f_n,(p(n)) \simeq (g_n,q(n))$.

Proof We begin with the observation that if K is finitely generated, then for any system (H_n,ψ_n),

(5.4) $$\mathrm{Hom}(K,H_\infty) = \underrightarrow{\lim}\ \mathrm{Hom}(K,H_n)$$

To see this note that given $f:K \longrightarrow H_\infty$, $f(K)$ is finitely generated. and thus there is an $n > 0$ and a finitely generated subgroup $K_n \subseteq H_n$ such that $\psi_{n\infty}(K_n) = f(K)$. Since

$$\ker(\psi_{n\ \infty}|_{K_n}) = \bigcup_K \ker(\psi_{n\ n+k}|_{K_n}).$$

is a subgroup of K_n, it must be finitely generated, and there exists a $k > 0$ such that

$$\ker(\psi_{n\infty}|_{K_n}) = \ker(\psi_{n\ n+k}|_{K_n}).$$

Replacing K_n by $\psi_{n\ n+k}(K_n)$, we may initially assume that θ_n =

$\psi_{n\infty}|_{K_n}:K_n \longrightarrow K$ is an isomorphism. Letting

$$g = \theta_n^{-1} f:K \longrightarrow H_n$$

we have that $\psi_{n\infty}\circ g = f$.

Let us suppose that there are two homomorphisms $f:K \longrightarrow H_n$, $g:K \longrightarrow H_m$ such that $\psi_{n\infty}\circ f = \psi_{m\infty}\circ g$. We may assume that $m = n$. Then $\psi_{n\infty}\circ(f - g) = 0$, i.e.,

$$(f - g)(K) \subseteq \bigcup_k \ker\psi_{n\ n+k}.$$

Since K is finitely generated, there is a $k \geq 0$ with $(f - g)(K) \subseteq \ker \psi_{n\ n+k}$, i.e., $\psi_{n\ n+k}\circ f = \psi_{n\ n+k}\circ g$, and we have proved (5.4).

Now suppose that we are given $h:G_\infty \longrightarrow H_\infty$. Then $h\circ\varphi_{1\infty}:G_1 \longrightarrow H_\infty$ may be used to find a map $h_1:G_1 \longrightarrow H_{n(1)}$ such that the diagram

$$\begin{array}{ccc} G_1 & \xrightarrow{\varphi_{1\infty}} & G_\infty \\ {\scriptstyle f_1}\downarrow & & \downarrow{\scriptstyle f} \\ H_{n(1)} & \xrightarrow[\psi_{1\infty}]{} & H_\infty \end{array}$$

commutes. In a similar manner we may find $m > n(1)$ and a homomorphism $f_2:G_2 \longrightarrow G_m$ such that the diagram

$$\begin{array}{ccc} G_2 & \xrightarrow{\varphi_{2\infty}} & G_\infty \\ {\scriptstyle f_2}\downarrow & & \downarrow{\scriptstyle f} \\ H_m & \xrightarrow[\psi_{m\infty}]{} & H_\infty \end{array}$$

commutes. We have that

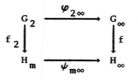

$$
\begin{aligned}
\psi_{m\,\infty}\circ(f_2\circ\varphi_1) &= f\circ\varphi_{2\,\infty}\circ\varphi_1\\
&= f\circ\varphi_{1\,\infty}\\
&= \psi_{1\,\infty}\circ f_1\\
&= \psi_{m\,\infty}\circ(\psi_{1\,m}\circ f_1)
\end{aligned}
$$

and thus for some $n(2) \geqslant m$,

$$(\psi_{m\;n(2)}\circ f_2\circ\varphi_1 = \psi_{1n(2)}\circ f_1.$$

Replacing f_2 by $\psi_{m\;n(2)}f_2$, we obtain a commutative diagram

$$
\begin{array}{ccccc}
G_1 & \longrightarrow & G_2 & \longrightarrow & G\\
\downarrow & & \downarrow & & \downarrow\\
H_{n(1)} & \longrightarrow & H_{n(2)} & \longrightarrow & H
\end{array}
$$

We continue by induction.

Finally say we are given $h = \varinjlim(f_n,p(n)) = \varinjlim(g_n,q(n))$. Composing the maps f_n, g_n with suitable ψ_{kh}, we obtain $f_n', g_n' : G_n \longrightarrow H_{r(n)}$ where $r(n) \geqslant p(n), q(n)$ and $(f_n', r(n)) \simeq (f_n, p(n))$, $(g_n', r(n)) \simeq (g_n, q(n))$. Since $\psi_{r(n)\infty}\circ(f_n' - g_n') = 0$ we have $\psi_{r(n)m}\circ(f_n' - g_n') = 0$ for sufficiently large m. Letting $f_n'' = \psi_{r(n)m}\circ f_n'$, $g_n'' = \psi_{r(n)m}\circ g_n'$, for an increasing sequence of m, we obtain $(f_n', r(n)) \simeq (f_n'', s(n)) = (g_n'', s(n)) \simeq (g_n', r(n))$. ■

Letting FA and CA be the categories of finitely genrated and countable abelian groups, respectively, we have that

$$\varinjlim : \text{inj}\ \underline{FA} \longrightarrow \underline{CA}$$

is an equivalence of categories, i.e.,

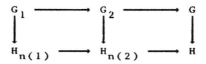

$$\underrightarrow{\lim}:\mathrm{Hom}((G_n,\varphi_n),(H_n,\psi_n)) \longrightarrow \mathrm{Hom}(G_\infty,H_\infty)$$

is a bijection, and each $H \in \underline{A}$ is isomorphic to $\underrightarrow{\lim}(G_n,\varphi_n)$ for some direct sequence (G_n,φ_n). We thus obtain a functor $S:\underline{CA} \longrightarrow$ inj $\underline{FA}$ with natural equivalences

$$S\circ\underrightarrow{\lim} \simeq \mathrm{id}, \qquad \underrightarrow{\lim}\circ S \simeq \mathrm{id}.$$

Specifically, for each H we let S(H) be a system (G_n, φ_n) with $\underrightarrow{\lim}(G_n,\varphi_n) \cong H$. Then for each homomorphism $f:H_1 \longrightarrow H_2$ we let $S(f) \in \mathrm{Hom}(S(H_1),p), (S(H_2),q))$ be the unique (to within equivalence) homomorphism with $\underrightarrow{\lim}\ S(f) = f$.

Inverse sequences of groups have an analogous inverse limit $\underleftarrow{\lim}$, and we have a functor

$$\underleftarrow{\lim}:\mathrm{pro}\underline{A} \longrightarrow \underline{A}$$

This functor is not as useful as $\underrightarrow{\lim}$ since, for example, the sequence

$$\cdots \xrightarrow{2} \mathbb{Z} \xrightarrow{2} \mathbb{Z}$$

is not equivalent to 0 in pro $\underline{A}$, even though $\underleftarrow{\lim}\ \mathbb{Z} = 0$. In the general case it is necessary to work with the objects in pro $\underline{A}$ themselves (i.e., "pro-groups") rather than their limits. Fortunately, we will see that an inverse system of finitely generated abelian groups may be split into inverse systems of torsion and torsion free finitely generated groups. The "*" and "d" dualities of §3 and §4 may then be used to convert these into direct systems of groups, which may then be classfied via Lemma 5.1.

It should be noted that the direct sequence analog of the following result is false since it would imply that for any countable abelian group G, $G \cong FG \oplus TG$ (see [R1] for a counterexample).

__Theorem 5.2__ Suppose that (G_n, φ_n) is an inverse system of finitely generated abelian groups. Then there is an isomorphism of pro-groups

$$(G_n, \varphi_n) \longrightarrow (FG_n \oplus Tg_n, F\varphi_n \oplus T\psi_n)$$

More precisely, there exist inverse sequence homomorphisms (θ_n, r_n), (η_n, s_n) for which the following diagrams commute

$$\begin{array}{ccccc} TG_{r(n)} & \longrightarrow & G_{r(n)} & \longrightarrow & FG_{r(n)} \\ {\scriptstyle T\varphi_{r(n)n}}\downarrow & & \downarrow{\scriptstyle \theta_n} & & \downarrow{\scriptstyle F\varphi_{r(n)n}} \\ TG_n & \longrightarrow & FG_n \oplus TG_n & \longrightarrow & FG_n \end{array}$$

$$\begin{array}{ccccc} TG_{s(n)} & \longrightarrow & FG_{s(n)} \oplus Tg_{s(n)} & \longrightarrow & FG_{s(n)} \\ {\scriptstyle T\eta_{s(n)n}}\downarrow & & \downarrow{\scriptstyle \eta_n} & & \downarrow{\scriptstyle F\eta_{s(n)n}} \\ TG_n & \longrightarrow & G_n & \longrightarrow & FG_{s(n)} \end{array}$$

and $\eta_n \circ \theta_{s(n)} = \varphi_{n\ r(s(n))}$, $\theta_n \circ \varphi_{r(n)} = \psi_{n\ s(r(n))}$.

Proof We let $T_n = T(G_n)$, $F_n = F(G_n)$ and we fix identifications $G_n \cong F_n \oplus T_n$. Then $\varphi_n : G_{n+1} \longrightarrow G_n$ has a corresponding representation

$$\varphi_n = \begin{pmatrix} a_n & 0 \\ b_n & c_n \end{pmatrix} \qquad a_n = F(\varphi_n),\ c_n = T(\varphi_n)$$

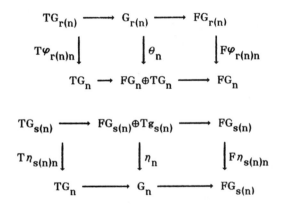

Similarly for $\varphi_{m,n}:G_m \longrightarrow G_n$, we have

$$\varphi_{mn} = \begin{pmatrix} a_{mn} & 0 \\ b_{mn} & c_{mn} \end{pmatrix} \qquad a_{mn} = F(\varphi_{mn}),\ c_{mn} = T(\varphi_{mn}).$$

Since the T_n are finite there is an isomorphic pro-group

$$\cdots \xrightarrow{c_2'} T_2' \xrightarrow{c_1'} T_1'$$

for which the c_n' are surjective. Thus there exists inverse sequence homomorphisms $f_n:T_{p(n)} \longrightarrow T_n'$, $g_n:T_{q(n)}' \longrightarrow T_n$ such that $g_nf_{q(n)} = c_{n,pq(n)}$, $f_ng_{p(n)} = c'_{n,qp(n)}$.

Consider the diagram

(5.5)

$$\begin{array}{ccc} F_{p(n+1)}\oplus T_{p(n+1)} & \xrightarrow{\begin{pmatrix} a_{p(n)p(n+1)} & 0 \\ b_{p(n)p(n+1)} & c_{p(n)p(n+1)} \end{pmatrix}} & F_{p(n)}\oplus T_{p(n)} \longrightarrow \\ \Big\downarrow \begin{pmatrix} a_{n+1\,p(n+1)} & 0 \\ h_{n+1} & f_{n+1} \end{pmatrix} & & \Big\downarrow \begin{pmatrix} a_{np(n)} & 0 \\ h_n & f_n \end{pmatrix} \\ F_{n+1}\oplus T'_{n+1} & \xrightarrow[\begin{pmatrix} a_{n+1} & 0 \\ 0 & c'_{n+1} \end{pmatrix}]{} & F_n\oplus T'_n \longrightarrow \end{array}$$

$$\begin{array}{c} \longrightarrow \ \cdots\cdots \ \longrightarrow F_1\oplus T_1 \\ \Big\downarrow \begin{pmatrix} 1 & 0 \\ h_1 & f_1 \end{pmatrix} \\ \longrightarrow \ \cdots\cdots \ \longrightarrow F_1\oplus T_1' \end{array}$$

where we define $h_n:F_{p(n)} \longrightarrow T_n'$ inductively as follows. We let $h_1 =$

0. Having defined h_n, we need a homomorphism $h_{n+1}:F_{p(n+1)} \longrightarrow T'_{n+1}$ such that the leftmost square in (5.5) commutes. Since (f_n) is a system homomorphism, we are given $c'_{n+1}f_{n+1} = f_n c_{p(n)p(n+1)}$. Thus it suffices to find an h_{n+1} satisfying

$$c'_{n+1}h_{n+1} = h_n a_x + f_n b_x, \qquad x = p(n)p(n+1),$$

i.e., we have a commutative diagram

$$\begin{array}{ccc} & & T'_{n+1} \\ & h_{n+1} & \Big\downarrow c'_{n+1} \\ F_{p(n+1)} & \xrightarrow{\;h_n a_x + f_n b_x\;} & T'_n \end{array}$$

Since $F_{p(n+1)}$ is free and c'_{n+1} is surjective, the existence of h_{n+1} is immediate.

Composing with the maps

$$a_{nq(n)} \oplus g_n:F_{q(n)} \oplus T_{q(n)}' \longrightarrow F_n \oplus T_n,$$

we obtain the diagram

$$\begin{array}{ccc} G_{r(n+1)} = F_{r(n+1)} \oplus T_{r(n+1)} & \longrightarrow & F_{r(n)} \oplus T_{r(n)} = G_{r(n)} \\ \theta_{n+1} = \begin{pmatrix} \bar{a}_{n+1} & 0 \\ \bar{b}_{n+1} & \bar{c}_{n+1} \end{pmatrix} \Big\downarrow & & \Big\downarrow \theta_n = \begin{pmatrix} \bar{a}_n & 0 \\ \bar{b}_n & \bar{c}_n \end{pmatrix} \\ F_{n+1} \oplus T_{n+1} & \longrightarrow & F_n \oplus T_n \end{array}$$

where $r(n) = p(q(n))$, $\bar{a}_n = a_{nr(n)}$, $\bar{b}_n = g_n h_{q(n)}$, and $\bar{c}_n = g_n f_{q(n)} = c_{r(n)}$. From the latter relations we have

$$T\theta_n = T\varphi_{nr(n)}, \quad F\theta_n = F\varphi_{nr(n)}.$$

Consider the diagram of pro-groups

$$\begin{array}{ccccccccc}
0 & \longrightarrow & (T_n) & \longrightarrow & (G_n) & \longrightarrow & (F_n) & \longrightarrow & 0 \\
 & & \downarrow \cong & & \downarrow (\theta_n) & & \downarrow \cong & & \\
0 & \longrightarrow & (T_n) & \longrightarrow & (F_n+T_n) & \longrightarrow & (F_n) & \longrightarrow & 0
\end{array} \tag{5.6}$$

we have that pro $\underline{A}$ is an abelian category (see [S1]) and thus the "five lemma" holds. Hence there is an inverse $(\eta_n):(F_n \oplus T_n) \longrightarrow (G_n)$ for which the diagram corresponding to (5.6) also commutes. ■

6. PROPER AO ALGEBRAS.

We let $\underline{O}_p$ denote the C^*-algebras isomorphic to $A = O_{\underline{a}}$ for some non-cyclic irreducible matrix $\underline{a}$, together with the proper homomorphisms $\varphi:O_{\underline{a}} \longrightarrow O_{\underline{b}}$. A C^*-algebra A is a <u>proper AO algebra</u> if $A = \varinjlim A_n$ for a direct system (1.1) in $\underline{O}_p$. We note that $\underline{O}_p$ is not itself a category because identity maps are not proper.

We define the <u>homotopy category</u>, $\underline{HO}_p$ to have as objects the C^*-algebras in $\underline{O}_p$, and morphisms the homotopy classes $[\varphi]$ of proper maps φ. This is a category because if we choose $\varphi:A \longrightarrow A$ with $K_0(\varphi) = \mathrm{id}$, $\mathrm{Ext}(\varphi) = \mathrm{id}$, then for any $\psi:A \longrightarrow B$ and $\theta:C \longrightarrow A$, we have $[\psi] = [\psi\varphi]$ and $[\theta] = [\theta\psi]$. The existence of φ follows from Theorem 4.9, or more simply, by letting $\varphi(a) = sas^*$, where $s:1 \simeq e < 1$ in A.

Applying the definitions of §5, we say that two diagrams (1.2) are <u>isomorphic</u> if they are equivalent objects in the category inj $\underline{HO}_p$. i.e., we have morphisms $[\varphi_n]:A_n \longrightarrow B_{p(n)}$, $[\psi_n]:B_n \longrightarrow A_{q(n)}$ such that the following diagrams homotopy commutes:

$$\begin{array}{ccc} A_n & \longrightarrow & A_{n+1} \\ \downarrow & & \downarrow \\ B_{p(n)} & \longrightarrow & B_{p(n+1)} \end{array} \qquad \begin{array}{ccc} B_n & \longrightarrow & B_{n+1} \\ \downarrow & & \downarrow \\ A_{q(n)} & \longrightarrow & A_{q(n+1)} \end{array}$$

$$A_n \longrightarrow \cdots \longrightarrow A_{q(n)} \longrightarrow \cdots \longrightarrow A_{qp(n)}$$

$$B_n \longrightarrow \cdots \longrightarrow B_{p(n)} \longrightarrow \cdots \longrightarrow B_{pq(n)}$$

It essentially follows from [EK1, Corollary 5.3] that if $A = \varinjlim A_n = \varinjlim A_n'$, then the systems (A_n) and (A_n') are isomorphic. To see this, it suffices to show that if φ: $O_{\underline{a}} \longrightarrow \overline{\vee B_n}$, and φ is proper, then $\varphi \cong \psi$: $A \longrightarrow B_n$ for some n. Replacing F by $F \oplus \mathbb{C}(1 - \Sigma e_i)$, the "rotation argument" in [EK1, Theorem 3.13] still works. We say that two proper AO algebras have the same <u>shape</u> if the corresponding systems (1.2) are equivalent.

<u>**Theorem 6.1**</u> Two proper AO algebras A and B have the same shape if and only if $K_i(A) \cong K_i(B)$ (i = 0,1).

Proof We let $A = \varinjlim(A_n, \varphi_n)$, $B = \varinjlim(B_n, \psi_n)$. Since $K_i(A) = \varinjlim K_i(A_n)$ (i = 0,1) and the groups $K_i(A_n)$ are finitely generated, we can conclude from Lemma 5.1 that $(K_i(A_n)) \cong (K_i(B_n))$ (i = 0,1). We let

$$f_n^i: K_i(A_n) \longrightarrow K_i(B_{p(n)}), \qquad g_n^i: K_i(B_n) \longrightarrow K_i(A_{q(n)})$$

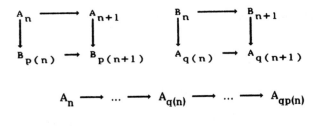

be direct system homomorphisms implementing these isomorphisms, i.e., letting $\varphi_{mn}^{i} = K_i(\varphi_{mn})$, we may assume that

$$(6.1) \qquad \begin{aligned} g^{i}_{p(n)} \circ f^{i}_{n} &= \varphi^{i}_{qp(n)n} \\ f^{i}_{q(n)} \circ g^{i}_{n} &= \psi^{1}_{pq(n)n} \end{aligned} \qquad (i = 0,1)$$

Using the natural isomorphisms $\mathrm{TExt}(A) \cong TK_0(A)^*$ and $\mathrm{FExt}(A) \cong K_1(A)^d$, we have corresponding inverse sequence maps

$$h_n = (f^1_n)^d \oplus T(f^0_n)^*: \mathrm{FExt}(B_{p(n)}) \oplus \mathrm{TExt}(B_{p(n)}) \longrightarrow \mathrm{FExt}(A_n) \oplus \mathrm{TExt}(A_n)$$

$$k_n = (g^1_n)^d \oplus T(g^0_n)^*: \mathrm{FExt}(A_{q(n)}) \oplus \mathrm{TExt}(A_{q(n)}) \longrightarrow \mathrm{FExt}(B_n) \oplus \mathrm{TExt}(B_n)$$

(note, the bonding maps are $(\psi^1_n)^d \oplus (T\psi^0_n)^*$, etc.). From Theorem 5.2 we have maps

$$\begin{array}{ccc} \mathrm{Ext}(B_{spr(n)}) & & \\ \downarrow \theta'_{pr(n)} & & \\ \mathrm{FExt}(B_{pr(n)}) \oplus \mathrm{TExt}(B_{pr(n)}) & \xrightarrow{h_{r(n)}} & \mathrm{FExt}(A_{r(n)}) \oplus \mathrm{TExt}(A_{r(n)}) \\ & & \downarrow \eta_n \\ & & \mathrm{Ext}(A_n) \end{array}$$

such that $T(\eta_n) = T(\varphi^0_{r(n)n})^*$, $T(\theta'_{pr(n)}) = T(\psi^0_{spr(n)\ \ pr(n)})^*$, and thus

$$F^e_n = \eta_n \circ h_{r(n)} \circ \theta'_{pr(n)}: \mathrm{Ext}(B_{spr(n)}) \longrightarrow \mathrm{Ext}(A_n)$$
$$F^0_n = \psi^0_{spr(n)\ pr(n)} \circ f^0_{r(n)} \circ \varphi^0_{r(n)n}: K_0(A_n) \longrightarrow K_0(B_{spr(n)})$$

satisfy $TF_n^e = (TF^0)^*$. From Theorem 4.9 we may find a proper homomorphism $F_n:A_n \longrightarrow B_{spr(n)}$ with $K_0(F_n) = F_n^0$ and $\mathrm{Ext}(F_n) = F_n^e$.

We similarly have a diagram

$$\begin{array}{ccc}
\mathrm{Ext}(A_{uqt(n)}) & & \\
\downarrow \theta_{qt(n)} & & \\
\mathrm{FExt}(A_{qt(n)}) \oplus \mathrm{TExt}(A_{qt(n)}) & \xrightarrow{k_{t(n)}} & \mathrm{FExt}(B_{t(n)}) \oplus \mathrm{TExt}(B_{t(n)}) \\
\downarrow \eta_n' & & \\
 & & \mathrm{Ext}(B_n)
\end{array}$$

such that if

$$G_n^e = \eta_n' \circ k_{t(n)} \circ \theta_{qt(n)}:\mathrm{Ext}(A_{uqt(n)}) \longrightarrow \mathrm{Ext}(B_n)$$

$$G_n^0 = \varphi^0_{uqt(n)\ qt(n)} \circ g^0_{t(n)n} \circ \psi^\circ_{t(n)n}:K_0(B_n) \longrightarrow K_0(A_{uqt(n)})$$

then $TG_n^e = (TG_n^0)^d$. We let $G_n:B_n \longrightarrow A_{uqt(n)}$ be such that $K_0(G_n) = G_n^0$, $\mathrm{Ext}(G_n) = G_n^e$.

Omitting some of the subscripts and letting $v(n) = uqtspr(n)$

$$K_0(G_{spr(n)} \circ F_n) = \varphi^0 \circ g^0 \circ \psi^0 \circ \psi^0 \circ f^0 \circ \varphi^0 = \varphi^0_{v(n)n},$$

$$\mathrm{Ext}(G_{spr(n)} \circ F_n) = \eta \circ h \circ \theta' \circ \eta' \circ k \circ \theta.$$

But we may assume that for suitable subscripts

$$\theta' \circ \eta' = (\psi^1)^d \oplus (T\psi^0)_*$$

and thus

$$h \circ \theta' \circ \eta' \circ k = (g^1 \circ \psi^1 \circ f^1)^d \oplus T(g^0 \circ \psi^0 \circ f^0)_*$$

$$= (\varphi^1)^d \oplus T(\varphi^0)_*$$

or, since θ and η' are sequence homomorphisms,

$$\mathrm{Ext}(G_{v(n)} \circ F_n) = \mathrm{Ext}(\varphi_{v(n)n}).$$

It follows that $G_{v(n)} \circ F_n \simeq \varphi_{v(n)n}$. Similarly, we obtain $F_{w(n)} \circ G_n \simeq \psi_{w(n)n}$ for suitable $w(n)$.

Using K_0 and Ext it is easy to check that the diagrams

$$\begin{array}{ccc} A_n & \xrightarrow{F_n} & B_{spr(n)} \\ \downarrow & & \downarrow \\ A_{n+1} & \xrightarrow[F_n]{} & B_{spr(n+1)} \end{array}$$

commute in the homotopy sense, i.e., (F_n) is a direct system morphism, and by symmetry, the same is true for (G_n). We thus conclude that $(A_n) \cong (B_n)$, i.e., A and B have the same shape. Since the converse is trivial, we are done. ■

REFERENCES

[B1] O. Bratteli. Inductive limits of finite dimensional C^*-algebras, Trans. Amer. Math. Soc. 171(1972), 195-234

[B2] L. G. Brown, Stable isomorphism of hereditary subalgebras of C^*-algebras, Pacific J. Math. 71(1977), 335-348

[B3] ———, The universal coefficient theorem for Ext and quasi-diagonality, in Operator Algebras and Group Representations-I, Proc. Internat. Conf. at Neptun, Romania, 1980, Monographs and Studies in Math., No. 17, Pitman, London, 1984

[B4] R. C. Busby. Double centralizers and extensions of C^*-algebras, Trans. Amer. Math. Soc. 132(1968), 79-99

[C1] J. Cuntz, On the homotopy groups of the space of endomorphisms of a C^*-algebra (with application to topological Markov chains), in the Operator Algebras and Group Representations-I, Proc. Internat. Conf. at Neptun, Romania, 1980, Monographs and Studies in Math., No. 17, Pitman, London, 1984, 124-137

[C2] ———, K-theory for certain C^*-algebras. Ann. of Math. (2)113 (1981), 181-197

[CK1] ——— and W. Krieger, A class of C^*-algebras and topological Markov chains, Invent. Math. 56(1980), 251-268

[D1] J. Dixmier, On some C^*-algebras considered by Glimm. J. Funct. Anal. 1(1967), 182-203

[E1] E. Effros, Dimensions and C^*-algebras, CBMS Regional Conference Series in Math., No. 46, AMS, Providence, R.I., 1981

[EK1] ———, and J. Kaminker, Homotopy continuity and shape theory for C^*-algebras, in Geometric Methods in

Operator Algebras, U.S. Japan Joint Seminar at Kyoto, 1983, Pittman, to appear

[E2] G. Elliott, On the classification of inductive limits of sequences of semi-simple finite dimensional algebras, J. Algebra 38(1976), 29-44

[E3] D. E. Evans, Gauge actions on O_A, J. Operator Theory, 7(1982), 79-100

[HS1] P. Hilton and U. Stammbach, A Course in Homological Algebra, Graduate Texts in Math., Vol. 4, Springe-Verlag, Berlin, 1971

[K1] G. G. Kasparov, The operator K-functor and extensions of C^*-algebras, Math. of the U.S.S.R. Izv, 16(1981), 513-572

[P1] W. L. Pachke, K-theory for actions of the circle group on C^*-algebras, J. Operator Theory 6(1981), 125-133

[R1] J. Rotman, The Theory of Groups: an introduction, Allyn and Bacon, Inc., Boston, 1965

[S1] S. Singh, Pro-homology and isomorphism of pro-groups, J. of Pure and Apl. Alg. 23(1982), 209-219

[S2] G. Skandalis, Some remarks on Kasparov theory, to appear

[T1] J. Taylor, Banach algebras and topology, in Algebras in Analysis (J. Williamson, ed.), Academic Press, New York, 1975

Addresses of authors

Edward G. Effros
Department of Mathematics
UCLA
Los Angeles, CA 90024

Jerome Kaminker
Mathematical Sciences Research Institute
2223 Fulton Stret
Berkeley, CA 94720

(after July, 1985)
Department of Mathematical Sciences
IUPUI
Indianapolis, IN 46223

SMALL UNITARY REPRESENTATIONS OF CLASSICAL GROUPS

Roger Howe

Dedicated to George W. Mackey

1. GENERAL DISCUSSION

In this paper, we will give a "natural" description of a certain subset of the unitary dual of $Sp_{2n}(\mathbb{R})$, the real symplectic group in 2n variables. The description will be in terms of the unitary duals of orthogonal groups. The representations of the symplectic group which we describe here are "small" in a well-defined sense explained below. The description relies heavily on the Mackey theory of induced representations, and on the theory of the oscillator representation. This paper is essentially a continuation of ↑H←. Results similar to those described here are valid for other classical Lie groups, and for classical groups over p-adic fields.

Let X be a finite-dimensional real vector space, with dual X^*. Set

$$W = X \oplus X^* \tag{1.1}$$

and define a symplectic form < , > on W by setting

$$\langle (x,\lambda),(x',\lambda')\rangle = \lambda(x') - \lambda'(x) \qquad x,x' \in X; \lambda,\lambda' \in X^* \tag{1.2}$$

Let Sp(W) = Sp be the group of isometries of the form < , >.

Let P ⋈ Sp be the (parabolic) subgroup of Sp consisting of elements which stabilize X inside W. Then

$$P \mid MN \tag{1.3}$$

where M is the Levi component of P and N is the unipotent radical. We may take M to be the subgroup of Sp which leaves both X and X^* invariant. Restricting the action of M to X gives us an isomorphism

(1.4) $$M \simeq GL(X)$$

Given $x,x'\in X$, we can form the symmetrized dyad

(1.5) $$E_{x,x'}(v) = \langle v,x\rangle x' + \langle v,x'\rangle x.$$

It is clear that $X \subseteq \ker E_{x,x'}$ and $\operatorname{im} E_{x,x'} \subseteq X$. It is easy to check that $1+E_{x,x'}$ preserves the form $\langle\ ,\ \rangle$. Let $S^2(X)$ be the second symmetric tensor power of X. Then the map $(x,x') \longrightarrow E_{x,x'}$ extends to an isomorphism

(1.6) $$S^2(X) \simeq N.$$

In particular N is abelian.

As is very well known, the Pontryagin dual $\hat{V}$ of a vector space V is isomorphic to its vector space dual V^* by the map

(1.7) $$\lambda(\cdot) \longrightarrow e^{2\pi i\lambda(\cdot)} \qquad \lambda\in V^*$$

In particular, in view of isomorphism (1.6), the Pontryagin dual of N is identifiable with the space $S^{2^*}(X)$ of symmetric bilinear forms on X.

Consider a unitary representation τ of Sp. We want to study how τ restricts to N. By the well-known representation theory of abelian groups, the restriction $\tau|N$ defines and is defined by a projection-valued measure on $\hat{N} \simeq S^{2^*}(X)$[M4].

If we first restrict τ to the parabolic subgroup P, then we obtain $\tau|N$ as the restriction of the representation $\tau|P$. Hence the projection-valued measure on $\hat{N}$ must be equivariant for the adjoint action Ad^*M of M on $\hat{N}$. We may identify this action with the standard action of GL(X) on $S^{2^*}(X)$. Thus there are only finitely many Ad^*M orbits in $\hat{N}$, corresponding to the isomorphism classes of symmetric bilinear forms on X. Such a class is described by its signature (p,q) and is isomorphic to a form on $\mathbb{R}^n$ (with n=dimX) having a matrix

$$\begin{bmatrix} 1_p & 0 & 0 \\ 0 & -1_q & 0 \\ 0 & 0 & 0 \end{bmatrix}$$

where 1_p is the pxp identity matrix. Hence p+q=r is the rank of this class of bilinear form.

Since our projection-valued measure on $S^{2*}(X)$ allows GL(X) as automorphisms, its restriction to any GL(X) orbit O will be absolutely continuous with respect to the unique GL(X) invariant measure class on O; and this restriction will have constant multiplicity. Hence the restricted representation $\tau | N$ has a very simple form: for each isomorphism type β of symmetric bilinear form on X we must specify the integer m_β which tells us the multiplicity of the spectral measure for $\tau | N$ on the GL(X) orbit O_β of forms of type β. The collection of integers m_β determines $\tau | N$.

The elements of $S^{2*}(X)$ having rank not more than a given number r form a subvariety B_r of codimension $\frac{(n-r)(n-r+1)}{2}$ in $S^{2*}(X)$. If a representation τ of Sp is such that the spectral measure of $\tau | N$ is concentrated on B_r we will say τ has <u>N-rank at most r</u>. If τ has N-rank at most r, and also $\tau | N$ has multiplicity zero on all orbits of B_{r-1}, we will say τ has <u>pure N-rank r</u>. If $\tau | N$ assigns multiplicity zero to all GL(X) orbits except for the orbit O_β for forms of type β, we will say τ is of <u>N-spectral type</u> β.

Because Sp(W) is not simply connected, and more particularly because its two-fold cover $\widetilde{Sp}$ plays an important role in the theory in connection with the oscillator representation, we want to consider representations of covering groups of Sp as well as of Sp itself. The above definitions concerning N-rank apply without change to any covering group of Sp.

In [H] the following result is proved.

Theorem 1.1. *Let* τ *be an irreducible unitary representation of some covering group of* Sp. *Then* τ *has pure* N-*rank* r *for some* $r \leq n = \dim X$. *If* $r<n$, *then* τ *has* N-*spectral type* β *for some bilinear form* β *of rank* r. *Also if* $r<n$, *then* τ *factors through* $\widetilde{Sp}$, *the* 2-*fold cover of* Sp; *and* τ *factors through* Sp *if and only if* r *is even.*

In view of this theorem, we will focus on describing $(\widetilde{Sp})^\wedge$, the unitary dual of Sp. Theorem 1.1 gives us a decomposition

$$(1.8) \qquad (\widetilde{Sp})^\wedge = \bigcup_{r\leq n} (\widetilde{Sp})^\wedge_r = (\widetilde{Sp})^\wedge_n \cup \Big(\bigcup_{r<n} \Big(\bigcup_{\operatorname{rank}\beta=r} (\widetilde{Sp})^\wedge_\beta \Big)\Big)$$

where $(\widetilde{Sp})^\wedge_r$ denotes the set of irreducible unitary representations of $\widetilde{Sp}$ of pure N-rank r, and $(\widetilde{Sp})^\wedge_\beta$ denotes the subset of $(\widetilde{Sp})^\wedge_r$ consisting of representations of N-spectral type β, for $\beta \in S^{2*}(X)$ of rank r.

Remark: When $r = \operatorname{rank} \beta$ is small enough, the subset $(\widetilde{Sp})^\wedge_\beta$ is both open and closed in $(\widetilde{Sp})^\wedge$. That is, $(\widetilde{Sp})^\wedge_\beta$ is isolated from the rest of $(\widetilde{Sp})^\wedge$. In [H] this is shown to be so when $r<(\frac{2n}{3})-2$, but this result is probably not sharp.

The representations in $(\widetilde{Sp})^\wedge_n$ constitute "most" of the representations of $\widetilde{Sp}$, and they are relatively large in the sense that their N-spectrum fills up an open set in $\hat{N}$. On the other hand the representations in $(\widetilde{Sp})^\wedge_\beta$ for some β of rank $r<n$ are small in the sense that their N-spectrum is concentrated on a subvariety of positive codimension in $\hat{N}$. Here we will be concerned with giving a description of the $(\widetilde{Sp})^\wedge_\beta$. The following result also proved in [H]

allows us to begin our description.

Let $X_1 \subseteq X$ be a subspace of dimension $r<n$. Let P_1 be the parabolic subgroup of Sp stabilizing X_1, and let $\widetilde{P}_1$ be the inverse image of P_1 in $\widetilde{Sp}$.

Theorem 1.2.

a) *Consider* $\tau \in (\widetilde{Sp})^{\wedge}_r$. *Then the restriction of* τ *to* $\widetilde{P}_1$ *is irreducible.*

b) *Given* $\tau, \sigma \in (\widetilde{Sp})^{\wedge}_r$, *if the restrictions of* τ *and* σ *to* $\widetilde{P}_1$ *are equivalent, then* τ *and* σ *are equivalent.*

This theorem tells us we have an injection $(\widetilde{Sp})^{\wedge}_r \longrightarrow (\widetilde{P}_1)^{\wedge}$. However, we can be more precise.

Let X_2 be a complement to X_1 in X, so

(1.9) $$X = X_1 \oplus X_2.$$

Set

(1.10) $$W_i = X_i \oplus X_i^*$$

so that

(1.11) $$W = W_1 \oplus W_2$$

is an orthogonal decomposition of W. We can write

(1.12) $$P_1 = M_1 N_1$$

where N_1 is the unipotent radical of P_1 and M_1 is a Levi component.

We can take M_1 to be the subgroup of P_1 which stabilizes X_1, X_1^* and W_2. Then restriction of elements of M_1 to $X_1 \oplus W_2$ yields an isomorphism

(1.13) $$M_1 \simeq GL(X_1) \times Sp(W_2).$$

The group N_1 is two-step nilpotent. Its center $Z(N_1)$ leaves $X_1 \oplus W_2$ pointwise fixed, hence belongs to $Sp(W_1)$. (We regard $Sp(W_1)$ as the subgroup of $Sp(W)$ leaving W_2 pointwise fixed.) Thus

(1.14) $$Z(N_1) = N_1 \cap Sp(W_1) = N \cap Sp(W_1) \simeq S^2(X_1).$$

The whole group N_1 acts as the identity on X_1 and on $(X_1 \oplus W_2)/X_1$. Thus the map

(1.15) $$n \longrightarrow 1-n$$

where 1 here denotes the identity map on W, defines a homomorphism from N_1 to $Hom(W_2, X_1)$. In fact the map (1.15) yields an isomorphism of $N_1/Z(N_1)$ with $Hom(W_2, X_1)$. Thus we have an exact sequence

(1.16) $$1 \longrightarrow S^2(X_1) \longrightarrow N_1 \longrightarrow Hom(W_2, X_1) \longrightarrow 1$$

Choose a non-degenerate symmetric bilinear form $\beta \in S^{2*}(X_1)$. When convenient we can extend β to a symmetric bilinear form of rank r on X by declaring X_2 to be the radical of the extended form. We denote by Ξ_β the character of $S^2(X_1) \simeq Z(N_1)$ or of $S^2(X) \simeq N$ associated to β.

Consider a representation $\tau \in (\widetilde{Sp})\hat{}_\beta$. It is easy to see that $\tau \mid \widetilde{Sp}(W_1)$ has pure $Z(N_1)$ spectral type β. It follows from the Mackey Imprimitivity Theorem [M4][Ri] that $\tau \mid \widetilde{P}_1$ is an induced representation, induced from a representation of the stabilizer of Ξ_β under the coadjoint action of P_1 on $Z(N_1)\hat{}$. Since $\widetilde{Sp}(W_2)$ centralizes $Z(N_1)$, as does N_1, we see that $\widetilde{Sp}(W_2)\cdot N_1$ is in this stabilizer. The stabilizer of Ξ_β in $GL(X_1)$ is simply O_β, the isometry group of β. Thus the full stabilizer in $\widetilde{P}_1$ of Ξ_β is

$$\widetilde{J}_\beta = (O_\beta \times Sp(W_2))^{\sim}\cdot N_1. \tag{1.17}$$

Let σ be the representation of $\widetilde{J}_\beta$ from which τ is induced. Then σ restricted to $Z(N_1)$ should be a multiple of Ξ_β. Let $A \subseteq Z(N_1)$ be the identity component of the kernel of Ξ_β. It is not difficult to check that, since β is a non-degenerate form on X_1, the quotient group N_1/A is a Heisenberg group. Thus there is a unique irreducible representation ρ_β of N_1 whose restriction to $Z(N_1)$ is a multiple of Ξ_β. Since $Sp(W_2)^{\sim}$ and $\widetilde{O}_\beta$ stabilize Ξ_β, they will also stabilize the representation ρ_β of N_1. From the theory of the oscillator representation [W1] we know that we can extend ρ_β to a representation ω_β of $(\widetilde{O}_\beta \times \widetilde{Sp}(W_2))\cdot N_1$.

Let ϵ denote the non-trivial element of the kernel of the projection map from $\widetilde{Sp}$ to Sp. Let ϵ_1 and ϵ_2 denote ϵ considered as an element of $\widetilde{O}_\beta$ and $\widetilde{Sp}(W_2)$ respectively. Then $\omega_\beta(\epsilon_1)$ and $\omega_\beta(\epsilon_2)$ will be unitary operators of order 2, and ω_β will factor from $\widetilde{O}_\beta \times \widetilde{Sp}(W_2)$ to $(O_\beta \times Sp(W_2))^{\sim}$ if and only if $\omega_\beta(\epsilon_1) = \omega_\beta(\epsilon_2)$. It can be computed that $\omega_\beta(\epsilon_1) = (-1)^{n-r}$

where n-r = $\dim X_2$, and that $\omega_\beta(\epsilon_2) = (-1)^r$ with r = $\dim X_1$. Thus ω_β factors to $\tilde{J}_\beta$ precisely when n is even. However there exists a character α of $\tilde{O}_\beta$ which takes the value -1 on ϵ_1. Hence if we tensor ω_β with one of these characters we can obtain a representation of $\tilde{J}_\beta$; but this procedure is slightly non-canonical, as the character of $\tilde{O}_\beta$ with which we must tensor is defined only up to a character of O_β.

In any case, we know, again from Mackey's work [M2] on representations of group extensions, that the general representation of $\tilde{J}_\beta$ which restricts to a multiple of Ξ_β on $Z(N_1)$ has the form $\mu\otimes\omega_\beta$ where μ is a representation of $\tilde{O}_\beta\times\tilde{S}p(W_2)$ such that the tensor product factors to $\tilde{J}_\beta$. If the representation is to be irreducible, then μ must be irreducible. This implies μ must factor: $\mu \simeq \mu_1\otimes\mu_2$, where μ_1 is an irreducible representation of $\tilde{O}_\beta$ and μ_2 is an irreducible representation of $\tilde{S}p(W_2)$.

The intersection $Sp(W_2)\cap N$ is non-trivial; in fact it is just $S^2(X_2) \subseteq S^2(X)$. By studying the effect of the $(Sp(W_2)\cap N)$-spectrum of μ_2 on the N-spectrum $\mu\otimes\omega_\beta$, we conclude that μ_2 must be trivial. Thus $\mu \simeq \mu_1\otimes 1$ is essentially just a representation of $\tilde{O}_\beta$. Then according to the discussion above, the condition that $\mu\otimes\omega_\beta$ factor to $\tilde{J}_\beta$ is simply that $\mu_1(\epsilon_1) = (-1)^n$. So if n = $\dim X$ is even, μ_1 is just a representation of O_β, while if n is odd, μ_1 is a twist by the character α of $\tilde{O}_\beta$ of a representation of O_β.

In summary, we have the following result, also from [H].

Theorem 1.3. *There is a natural injective mapping*

(1.18)
$$\delta_\beta: (\widetilde{Sp})^{\widehat{}}_\beta \longrightarrow \widehat{O}_\beta$$

defined by

(1.19)
$$\text{ind}_{\widetilde{J}_\beta}^{\widetilde{P}_1}(\delta_\beta(\tau)\alpha^n \otimes 1)\otimes\omega_\beta \simeq \tau \mid \widetilde{P}_1$$

for $\tau \in (\widetilde{Sp})^{\widehat{}}_\beta$. *Here* α *is a character of* $\widetilde{O}_\beta$ *non-trivial on* ϵ_1, *the tensor product in parentheses is an outer tensor product of* $\delta_\beta(\tau)\alpha^n \in (O_\beta)^{\widehat{}}$ *and* $1 \in (\widetilde{Sp}(W_2))^{\widehat{}}$, *and the tensor product with* ω_β *is an inner tensor product of representations of* $(\widetilde{O}_\beta \times \widetilde{Sp}(W_2)) \cdot N_1$.

Given Theorem 1.3 an obvious question to ask is, what is the range of δ_β? This question was left open in [H]. The main result of the present paper is an answer to this question.

Theorem 1.4. *If* $r = \text{rank}\,\beta = \dim X_1 \leq \dim X-2$, *then the map* δ_β *is surjective, i.e., it gives a bijection between* $(\widetilde{Sp})^{\widehat{}}_\beta$ *and* $\widehat{O}_\beta$.

Remarks: a) There is no particular reason to expect this result to fail when r=n-1. However the argument we give here does not apply to that case, for reasons that will be clear.

b) W. Rossmann has proved a similar result (oral communication).

c) If one believes a similar theory holds for representations of orthogonal groups, and for complex groups, it becomes possible to construct an interesting set of isolated points of $(Sp_{2n}(\mathbb{C}))^{\widehat{}}$. Namely for $O_r(\mathbb{C})$, for r=1 and r$\geq$5, the two one-dimensional representations are isolated in the unitary dual. When n$\geq$r+2, the one-dimensional representations of $O_r(\mathbb{C})$ will give rise to representations of $Sp_{2n}(\mathbb{C})$. And when r is in the appropriate range and at least when $2n \geqslant 3r+6$, the representations of $Sp_{2n}(\mathbb{C})$ that result will be isolated in its unitary dual. Then by the reverse process, these representations will give rise to isolated points in $(O_r(\mathbb{C}))\hat{}$ for r sufficiently large, and these representations of $O_r(\mathbb{C})$ will in turn yield isolated points in $Sp_{2n}(\mathbb{C})$ for n large enough, and so on back and forth. Thus as $n \longrightarrow \infty$ we obtain a finite but large collection of isolated points in the unitary dual of $Sp_{2n}(\mathbb{C})$. All these representations should be "unipotent" in the sense of Vogan.

d) There is some hope that the theory developed here for small representations has an extension which would allow one to develop a systematic picture of the whole unitary dual of classical groups. Such a picture would probably be complementary to the one being developed through the Langlands-Vogan classification theory.

The proof of Theorem 1.4 is given in the next section. It consists primarily in reversing the analysis, described more fully in [H], that culminates in Theorem 1.3. Obviously from a representation $\mu \in \hat{O}_\beta$, we can construct a representation of $\tilde{P}_1$ by putting μ where $\delta_\beta(\tau)$ is in the left hand side of equation (1.19). The problem then is to show that this representation of $\tilde{P}_1$ extends to a representation of $\widetilde{Sp}$. This is done by describing the extension explicitly on the parabolic subgroup $\tilde{P}$ and using the fact that $\tilde{P}_1$ and $\tilde{P}$ together generate $\widetilde{Sp}$.

2. PROOF OF THEOREM 1.4

This section is devoted to proving Theorem 1.4. The proof is not especially difficult, but it requires some courage. It is written so as to give as explicit a description as I know for the representation. We begin by reestablishing notation.

Consider two vector spaces X_1 and X_2 and set

(2.1)
$$X = X_1 \oplus X_2.$$

Let $X^* = X_1^* \oplus X_2^*$ be the vector space dual to X. Equip

(2.2)
$$W = X \oplus X^*$$

with the canonical symplectic form

(2.3)
$$\langle(x,\lambda),(x',\lambda')\rangle = \lambda(x') - \lambda'(x).$$

Let $Sp(W) = Sp$ be the group of isometries of the form given in (2.3) and let $\mathfrak{sp}(W) = \mathfrak{sp}$ be its Lie algebra. If we decompose $\mathfrak{sp}$ according to the decomposition of $W = X_1 \oplus X_2 \oplus X_2^* \oplus X_1^*$, we get

(2.4)
$$\mathfrak{sp} = \begin{bmatrix} A_1 & T_{12} & S_{12} & B_1 \\ T_{21} & A_2 & B_2 & S_{12}^* \\ R_{12} & C_2 & -A_2^* & -T_{12}^* \\ C_1 & R_{12}^* & -T_{21}^* & -A_1^* \end{bmatrix}$$

where the various pieces of this matrix can be identified as follows.

i) $A_i \in \mathrm{End}(X_i)$, and $A_i^* \in \mathrm{End}(X_i^*)$ is the adjoint of A_i.

ii) $T_{ij} \in \mathrm{Hom}(X_j,X_i)$, and $T_{ij}^* \in \mathrm{End}(X_i^*,X_j^*)$ is the adjoint of T_{ij}.

iii) $B_i \in S^2(X_i)$, the symmetric product of X_i with itself. Note according to its place in the matrix, we should have $B_i \in \mathrm{Hom}(X_i^*,X_i)$; but this space can be identified with $(X_i^* \otimes X_i^*)^*$, the space of bilinear forms on X_i^*, or with $X \otimes X$. Under these identifications B_i is symmetric.

iv) $S_{12} \in Hom(X_2^*, X_1) \simeq X_1 \otimes X_2$; and $S_{21}^* \in Hom(X_1^*, X_2)$ is its adjoint.

v) $C_i \in S^2(X_i^*)$ and $R_{ij} \in Hom(X_1, X_2^*)$, with comments as in iii) and iv).

As usual we let $\widetilde{Sp}(W) = \widetilde{Sp}$ denote the 2-fold cover of Sp. We will construct some representations of Sp if $\dim X_1$ is even, and of $\widetilde{Sp}$ if $\dim X_1$ is odd. Our construction will proceed in stages, as we define the representation on the various pieces of $\widetilde{Sp}$ corresponding to the decomposition of sp given in (2.4). To simplify slightly the discussion, we will explicitly deal only with the case when $\dim X_1$ is even.

Fix an element β of $S^{2*}(X_1)$ - a symmetric bilinear form on X_1. Let $O_\beta = O$ denote the group of isometries of β. Let σ be an irreducible unitary representation of O. Set $G_1 = GL(X_1) \times Hom(X_2, X_1)$, and consider the induced representation

$$\rho = \mathrm{ind}_O^{G_1} \sigma. \tag{2.5}$$

Let H be the Hilbert space on which σ is realized. Then the space of ρ can be taken to be the space of functions $f: G_1 \longrightarrow H$ satisfying

$$f(rg) = \sigma(r)f(g) \qquad r \in O, g \in G_1. \tag{2.6}$$

The action of G_1 is as usual, by right translation:

$$\rho(g')f(g) = f(gg') \tag{2.7}$$

We can use the semidirect product structure of G_1 to describe ρ in a slightly more explicit fashion. We write a typical element of G_1 as $g_1 = ax$ with $a \in GL(X_1)$ and $x \in Hom(X_2,X_1)$. Then if $r \in O$, we have $rg_1 = (ra)x$. Therefore we can realize the space of ρ as the space Y of functions $f: GL(X_1) \times Hom(X_2,X_1) \longrightarrow H$ satisfying

$$(2.6)' \qquad f(ra,x) = \sigma(r)f(a,x) \qquad r \in O, a \in GL(X_1), x \in Hom(X_2,X_1).$$

When we use the format of formula (2.6)' the action of G_1 is described by the rules

$$(2.7)' \qquad \begin{aligned} &a) \ \rho(a')f(a,x) = f(aa',a'^{-1}x) && a,a' \in GL(X_1) \\ &b) \ \rho(x')f(a,x) = f(a,x+x') && x,x' \in Hom(X_2,X_1). \end{aligned}$$

In formula b) we have used + to indicate the group law in $Hom(X_2,X_1)$.

Remarks: At this point we should make explicit a feature of our notation that might, unexplained, cause confusion. On the one hand, we are identifying $Hom(X_2,X_1)$ and $GL(X_1)$, and so forth, with subgroups of $GL(X)$. This means that $a \in GL(X_1)$ is taken to stand for the matrix $\begin{bmatrix} a & 0 \\ 0 & 1_2 \end{bmatrix}$ in $GL(X)$, where 1_2 is the identity operator on X_2; and likewise $x \in Hom(X_2,X_1)$ is proxy for the matrix $\begin{bmatrix} 1_1 & x \\ 0 & 1_2 \end{bmatrix}$ in $GL(X)$.

If we take the product ax as elements of $GL(X)$, we get the matrix $\begin{bmatrix} a & ax \\ 0 & 1_2 \end{bmatrix}$ where the upper right entry is again in $Hom(X_2,X_1)$ and is the natural product of an element of $End(X_1)$ with an element of $Hom(X_2,X_1)$, and is again an element of $Hom(X_2,X_1)$. In other words, our notation does not distinguish whether ax is intended to be a product in GL(X), or a product between $End(X_1)$ and $Hom(X_2,X_1)$. Which one is meant will I hope be clear from context. In the setting up of formula (2.6)', it is group multiplication in GL(X) that is involved, but on the right hand side of formula (2.7)'a), multiplication between $End(X_1)$ and $Hom(X_2,X_1)$ is involved.

The representation of Sp we will construct will act on the space Y described by equations (2.6) or (2.6)'. We will construct stepwise an extension of ρ to larger subgroups of Sp. Besides being explicit, this construction emphasizes the smallness of these representations of Sp.

There is a natural action of $GL(X_1)$ on $S^{2*}(X_1)$ by precomposition: if γ is a symmetric bilinear form on X_1 and $a \in GL(X_1)$, then

$$(2.8) \qquad a(\gamma)(v,v') = \gamma(a^{-1}(v),a^{-1}(v')).$$

In particular, given our form β we can produce the forms $a(\beta)$ for $a \in GL(X_1)$.

The space $S^{2*}(X_1)$ is naturally dual to the space $S^2(X_1)$ which occupies the upper right hand corner of the matrix (2.4). By the usual yoga, we can also identify $S^{2*}(X_1)$ with the Pontrjagin dual of $S^2(X_1)$. Thus each element γ of $S^{2*}(X_1)$ corresponds to a character Ξ_γ of $S^2(X_1)$. Explicitly, we have

$$(2.9) \qquad \Xi_\gamma(b) = e^{2\pi i\gamma(b)} \qquad \gamma \in S^{2*}(X_1), b \in S^2(X_1).$$

Let us define an action of $S^2(X_1)$ on the space Y by

$$(2.10) \qquad \rho(b)f(a,x) = \Xi_{a^{-1}(\beta)}(b)f(a,x).$$

It is clear by inspection of formulas (2.10) and (2.7)' that $\rho(S^2(X_1))$ and $\rho(\mathrm{Hom}(X_2,X_1))$ commute. It is easy to compute using these formulas that

$$\rho(a)\rho(b)\rho(a^{-1}) = \rho(a(b)).$$

Hence formulas (2.10) and (2.7)' in fact define a representation ρ of the semidirect product

$$G_2 = GL(X_1)\ltimes(\mathrm{Hom}(X_2,X_1)\oplus S^2(X_1)).$$

In fact it is easy to check that ρ is still an induced representation. The action of O on $S^{2*}(X_1)$ preserves β by definition of O. Hence we can extend the action σ of O on the space H to an action $\sigma\otimes\Xi_\beta$ of $O\ltimes S^2(X_1)$. The reader may verify that our representation ρ is now the representation of G_2 induced from $O\ltimes S^2(X_1)$ by $\sigma\otimes\Xi_\beta$.

Next consider the group $X_1\otimes X_2$ which accounts for the entry S_{12} of the matrix in (2.4). We have canonical isomorphisms

$$\mathrm{Hom}(X_2^*,X_1) \simeq X_1\otimes X_2 \simeq \mathrm{Hom}(X_1^*,X_2).$$

Given $y\in X_1\otimes X_2$, we will denote also by y the equivalent element in $\mathrm{Hom}(X_2^*,X_1)$ and by y^* its adjoint in $\mathrm{Hom}(X_1^*,X_2)$. Similarly given $x\in\mathrm{Hom}(X_2,X_1)$ we let x^* be its adjoint in $\mathrm{Hom}(X_1^*,X_2^*)$. We can then form products xy^* and yx^* in $\mathrm{Hom}(X_1^*,X_1) \simeq X_1\otimes X_1$. There is a natural projection

$$p\colon X_1\otimes X_1 \longrightarrow S^2(X_1)$$

and we have $p(xy^*) = p(yx^*) = \frac{1}{2}(xy^*+yx^*)$. By composing with p, we may regard characters of $S^2(X_1)$ as characters of $X_1 \otimes X_1$. We set

$$(2.11) \qquad \rho(y)f(a,x) = \Xi_{a^{-1}(\beta)}(xy^*)f(a,x).$$

Direct computation using formulas (2.7)', (2.10), and (2.11) shows that

$$(2.12) \qquad \text{a)} \ \rho(a)\rho(y)\rho(a)^{-1} = \rho(ay) \qquad a \in GL(X_1), y \in X_1 \otimes X_2$$

$$\text{b)} \ \rho(x)\rho(y)\rho(x^{-1})\rho(y^{-1}) = \rho(p(xy^*)).$$

The subgroup N_1 of Sp whose Lie algebra is spanned by $Hom(X_2,X_1)$, $X_2 \otimes X_1$, and $S^2(X_1)$ is two-step nilpotent. It fits in an exact sequence

$$(2.13) \qquad 1 \longrightarrow Hom((X_2 \oplus X_2^*),X_1) \longrightarrow N_1 \longrightarrow S^2(X_1) \longrightarrow 1.$$

The relations (2.12) show that formula (2.11) fits with the earlier formulas for ρ to define a representation ρ of $G_3 = GL(X_1) \ltimes N_1$. Again this ρ is an induced representation. If we extend the character Ξ_β from $S^2(X_1)$ to $(X_1 \otimes X_2) \oplus S^2(X_1)$ by letting Ξ_β be trivial on $X_1 \otimes X_2$, then ρ is induced from the representation of $\sigma \otimes \Xi_\beta$ of $O \ltimes (X_1 \otimes X_2 \oplus S^2(X_1))$.

Remarks: It is worth noting that we now have an irreducible representation. Indeed since the representation ρ of G_3 is an induced representation, we can apply Mackey's criterion [M4] for irreducibility of an induced representation. But the group $(X_1 \otimes X_2) \oplus S^2(X_1)$ is normal in G_3, and the stabilizer in $GL(X_1) \ltimes Hom(X_2,X_1)$ of the character Ξ_β is exactly O. Hence the irreducibility criterion applies to guarantee ρ is irreducible.

Next consider the group $S^2(X_2)$. Just as for X_1, we may consider $S^{2*}(X_2)$ to be the Pontrjagin dual of $S^2(X_2)$. Thus, given an inner product γ on X_2, we have an associated character Ξ_γ of $S^2(X_2)$, given as in formula (2.9). From an inner product $\gamma_1 \in S^{2*}(X_1)$ and a map $x \in \mathrm{Hom}(X_2, X_1)$ we can compose γ_1 with x to produce an inner product $x^*(\gamma_1) = \gamma_1 \circ x$ on X_2. We use this procedure to define an action of $S^2(X_2)$ on Y.

$$\rho(b_2)f(a,x) = \Xi_{\beta \circ a \circ x}(b_2)f(a,x) \qquad b_2 \in S^2(X_2). \tag{2.14}$$

Here β is our initial quadratic form, $a \in GL(X_1)$, and $x \in \mathrm{Hom}(X_2, X_1)$.

Since $\rho(S^2(X_1))$, $\rho(S^2(X_2))$, and $\rho(X_1 \otimes X_2)$ all consist of multiplication operators, they clearly commute with one another. Thus the actions of these three groups fit together to define an action of the abelian group $S^2(X) = S^2(X_1 \oplus X_2) = S^2(X_1) \oplus S^2(X_2) \oplus (X_1 \otimes X_2)$, which is the subgroup of Sp corresponding to the upper right hand quarter of the matrix in (2.4). We can extend the character Ξ_β from $S^2(X_1)$ to $S^2(X)$ by letting it be trivial on $X_1 \otimes X_2$ and on $S^2(X_2)$. This is equivalent to extending β from X_1 to X by declaring X_2 to be in the radical of β, then taking the associated character of $S^2(X)$. It is clear that the extended character will be invariant under the action of O on $S^2(X)$, so we can have the representation $\sigma \otimes \Xi_\beta$ of $O \ltimes S^2(X)$. One sees that formula (2.14) combined with the earlier formulas (2.11) and (2.7)' gives us the representation of

$$G_4 = (GL(X_1) \ltimes \mathrm{Hom}(X_2, X_1)) \ltimes S^2(X) \subset G_3 \cdot S^2(X_2)$$

induced from $O \ltimes S^2(X)$.

It is quite easy to add to this action an action of $GL(X_2)$, although care must be taken to add the right one. We have assumed $\dim X_1$ is even, say $\dim X_1 = 2m$. Then the inner product β on X_1 will have signature (p,q) with $p+q=2m$. Hence $p-q = 2(m-q)$. Set

$$\rho(a_2)f(a,x) = (\det a_2)^{m-q} |\det a_2|^{q} f(a,xa_2) \tag{2.15}$$

We leave it to the reader to check that (2.15) fits together with the previous formulas defining ρ to yield a representation of the group

$$G_5 = GL(X_2) \ltimes G_4$$

which is a parabolic subgroup of Sp, namely the stabilizer of the isotropic spaces X_1 and X. In fact the representation ρ is still an induced representation. The action of $GL(X_2)$ on $S^2(X)$ clearly preserves $S^2(X_2)$ and $X_1 \otimes X_2$, and leaves $S^2(X_1)$ fixed pointwise. Therefore the character Ξ_β of $S^2(X)$ is fixed under conjugation by $GL(X_2)$. Also $GL(X_2)$ commutes with $GL(X_1)$ inside G_5, and in particular commutes with O. Therefore if ψ is any character of $GL(X_2)$ we may form the representation $\sigma \otimes \psi \otimes \Xi_\beta$ of $(O \times GL(X_2)) \ltimes S^2(X)$. The representation ρ of G_5 is induced from $\sigma \otimes \psi \otimes \Xi_\beta$ with

$$\psi(a) = (\operatorname{sgn} \det a_2)^{m-q}. \tag{2.16}$$

To extend the representation ρ beyond G_5 is more difficult than what we have done so far. First we will extend it to the parabolic subgroup P of Sp which stabilizes X. This group P is isomorphic to $GL(X) \ltimes S^2(X)$. The Lie algebra of P consists of matrices as in (2.4) whose entries in the lower left quarter (R_{12}, C_2, C_1, R_{12}^*) are zero. To do this we observe that since our inner product β, considered as an inner product on X, has radical X_2, it will be invariant under the subgroup $GL(X_2)\ltimes Hom(X_1,X_2)$ of GL(X). Thus β will be fixed by $(O\ltimes GL(X_2))\ltimes Hom(X_1,X_2)$. We can therefore extend the representation $\sigma\otimes\psi\otimes\Xi_\beta$ of $(O\times GL(X_2))\ltimes S^2(X)$ to $((O\times GL(X_2))\ltimes(Hom(X_1,X_2))\ltimes S^2(X)$ by letting $Hom(X_1,X_2)$ act trivially. Do so, and consider the representation ρ of P induced (unitarily) from this extended $\sigma\otimes\psi\otimes\Xi_\beta$. By the Bruhat decomposition the double coset

$$((GL(X_1)\times GL(X_2))\ltimes Hom(X_1,X_2))\cdot Hom(X_2,X_1)$$

is open dense and of full measure in GL(X). Since

$$(((O\times GL(X_2))\ltimes Hom(X_1,X_2))\ltimes S^2(X))\cap G_5 = (O\times GL(X_2))\ltimes S^2(X)$$

it follows from the Mackey decomposition theorem [M2] that the restriction of ρ to G_5 is just the representation induced from $\sigma\otimes\psi\otimes\Xi_\beta$ on $(O\times GL(X_2))\ltimes S^2(X)$; in other words, this induced representation of P is an extension of the ρ we have so far defined on G_5. Using the general formulas (2.6) and (2.7) for induced representations, and taking into account our coordinates, as described in (2.6)' and (2.7)', we can compute that for $y\in Hom(X_1,X_2)$

$$(2.17) \qquad \rho(y)f(a,x) = \psi(1+yx)\,|\det(1+yx)|^{-m}f(a(1+xy),(1+xy)^{-1}x).$$

Remarks: Again I caution the reader about the possible notational pitfalls. On the left hand side of (2.17), the y in $\rho(y)$ stands for the element of GL(X) to which $y\in Hom(X_1,X_2)$ is identified. However, on the right hand side, the product xy is just the usual product between $x \in Hom(X_2,X_1)$ and $y \in Hom(X_1,X_2)$, so that $xy \in End(X_1)$, and also $1+xy \in End(X_1)$ (and for generic x, $1+xy \in GL(X_1)$). So det in det(1+yx) is the determinant function on $End(X_2)$.

Next we will extend ρ from G_5 to P_1, the parabolic subgroup of Sp consisting of maps stabilizing X_1. We have observed that on G_3, the representation ρ is induced from the subgroup $O \ltimes (X_1 \otimes X_2 \oplus S^2(X_1))$, from the representation $\sigma \otimes \Xi_\beta$. We have

$$O \ltimes (X_1 \otimes X_2 \oplus S^2(X_1)) \subseteq O \ltimes N_1 \subseteq G_3.$$

Here N_1 is as described in the sequence (2.13). Using induction in stages we can first induce up to $O \ltimes N_1$, then up to all of G_3.

Let τ denote the representation of $O \ltimes N_1$ induced from $\sigma \otimes \Xi_\beta$. As we have seen, the group N_1 is two-step nilpotent with center $S^2(X_1)$. Since the character Ξ_β of $S^2(X_1)$ is invariant under conjugation by O, the representation τ of $O \ltimes N_1$ will just be a multiple of Ξ_β on $S^2(X_1)$. Let A be identity component of the kernel of Ξ_β on $S^2(X_1)$. Then τ factors to a representation of $O \ltimes (N_1/A)$.

Since β is a non-degenerate inner product on X_1, the group N_1/A is a Heisenberg group. Hence there is a unique irreducible representation ρ_β of N_1 with central character Ξ_β. Since N_1 is normalized by $O \times Sp(W_2)$, where $W_2 = X_2 \oplus X_2^*$, the results of Shale [Sh] and Weil [W1] tell us we can extend ρ_β to some covering group of $(O \times Sp(W_2)) \ltimes N_1$. Since we have assumed $\dim X_1$ is even, we can in fact extend ρ_β to an $(O \times Sp(W_2)) \ltimes N_1$ itself -- we don't need a covering group. This extension is uniquely defined up to a character of O. We will specify which one we want in a moment.

Let us call the extension ω_β.

The group $(X_1 \otimes X_2) \oplus S^2(X_1)$ is maximal abelian in N_1, and $(X_1 \otimes X_2) \oplus (S^2(X_1)/\Lambda)$ is a maximal abelian subgroup in N_1/A. Let Ξ_β now denote the extension of Ξ_β on $S^2(X_1)$ to $(X_1 \otimes X_2) \oplus S^2(X_1)$ which is trivial on $X_1 \otimes X_2$. It is well-known [W1] that the representation ρ_β of N_1 may be realized as the representation induced from Ξ_β. More explicitly we can realize ρ_β on $L^2(\mathrm{Hom}(X_2, X_1))$ by the formulas

$$
\begin{array}{llll}
(2.18) & a) & \rho_\beta(x')f(x) = f(x+x') & x, x' \in \mathrm{Hom}(X_2, X_1) \\
 & b) & \rho_\beta(y)f(x) = \Xi_\beta(xy^*)f(x) & y \in X_1 \otimes X_2 \\
 & c) & \rho_\beta(b)f(x) = \Xi_\beta(b)f(x) & b \in S^2(X_1)
\end{array}
$$

In these formulas, f is a function on $\mathrm{Hom}(X_2, X_1)$. From [Sa] or [Ra] we can then find that the representation ω_β on $O \times Sp(W)$ can be taken to satisfy the following formulas

$$
\begin{array}{llll}
(2.19) & a) & \omega_\beta(r)f(x) = f(r^{-1}x) & r \in O \\
 & b) & \omega_\beta(b_2)f(x) = \Xi_{\beta \circ x}(b_2)f(x) & b_2 \in S^2(X_2) \\
 & c) & \omega_\beta(a_2)f(x) = \psi(a_2)|\det a_2|^m f(xa_2) & a_2 \in GL(X_2)
\end{array}
$$

Since $O \simeq ((O \times Sp(W_2)) \ltimes N_1)/(Sp(W_2 \ltimes N_1))$, we may consider the representation σ of O to be a representation of $(O \times Sp(W_2)) \ltimes N_1$. Hence we may form the (inner) tensor product representation $\sigma \otimes \omega_\beta$. A comparison of formulas (2.18) and (2.19) with formulas (2.6)', (2.7)', (2.10), (2.11), and (2.15) shows that the representation ρ of G_5 is just the restriction to G_5 of the representation of G_7 induced from

$\sigma \otimes \omega_\beta$ on $(O \times Sp(W_2)) \ltimes N_1$. In other words, we can extend ρ to P_1 by means of ω_β. Following [Sa] or [Ra] we may give a formula for the action of certain elements of $Sp(W_2)$. For example, consider a decomposition

$$X_2 = L \oplus C$$

where L is a line and C a complementary hyperplane. This yields a similar decomposition

$$X_2^* = L^* \oplus C^*$$

of dual spaces. Choose a vector $v \in L$, and let v^* be the element of L^* such that $v^*(v)=1$. Let s denote the element of $Sp(W_2)$ such that

$$s(v) = v^* \qquad s(v^*) = -v$$

and s acts as the identity on $C \oplus C^*$. Write

$$\mathrm{Hom}(X_2,X_1) = \mathrm{Hom}(C,X_1) \oplus \mathrm{Hom}(L,X_1).$$

Given $u \in X_1$, let $x_u \in \mathrm{Hom}(L,X_1)$ be the map which sends v to u. Then

$$\rho(s)(f)(a,x) = c\,|\det a| \int_{X_1} f(a,x+x_u)\Xi(\beta(ax(v),a(u)))du \tag{2.20}$$

where c is a number depending on v, β and the choice of Haar measure du on X_1. Also $\Xi(t) = e^{2\pi it}$ as usual.

We have now described ρ on the maximal parabolic subgroups P and P_1 of Sp(W). These two groups generate Sp(W). Hence if there is a representation ρ of Sp(W) which restricts to the ρ's defined above on P and P_1, this ρ is clearly uniquely determined. Thus if ρ

exists we have in principle described it. So our main worry now is about the existence of ρ.

In any event, we do have a representation of each of the groups P and P_1, and these representations have a common restriction to $G_5 = P \cap P_1$. Let G_8 be the group of unitary operators generated by $\rho(P)$ and $\rho(P_1)$. Then we certainly have a unitary representation of G_8; our question may be reformulated, is $G_8 \simeq Sp(W)$, or is it something strange?

To answer this question, we appeal to the theory of B-N pairs [T]. Before getting into the argument, let us make an observation about the ρ's so far defined. From formulas (2.17) and (2.20) we can compute directly that if $y \in Hom(X_1, X_2)$ has image in C, so that $x_u y = 0$, then $\rho(y)$ and $\rho(s)$ commute. This has the following implication. Let

$$X_2 = X_3 \oplus X_4 \tag{2.21}$$

be a decomposition of X_2 and set

$$X_5 = X_1 \oplus X_3. \tag{2.22}$$

Then $GL(X_5) \subseteq GL(X)$, and if $W_4 = X_4 \oplus X_4^*$ then $Sp(W_4) \subseteq Sp(W_2)$. Hence ρ is defined on $GL(X_5)$ by restriction from P and on $Sp(W_4)$ by restriction from P_1. I claim $\rho(GL(X_5))$ and $\rho(Sp(W_4))$ commute. To see this, observe that $GL(X_5)$ is generated by $GL(X_1)$, $GL(X_3)$, $Hom(X_3, X_1)$ and Hom (X_1, X_3). Of these, the only piece that is not contained in P_1 along with $Sp(W_4)$ is $Hom(X_1, X_3)$. Since ρ defines a representation of P_1, it will suffice to check that $\rho(Hom(X_1, X_3))$ commutes with $\rho(Sp(W_4))$. But $Sp(W_4)$ is generated by $Sp(W_4) \cap P$ and any other element, for example, by an element s as in formula (2.20) such that the decomposition $X_2 = L \oplus C$ involved in defining s satisfies

$$L \subseteq X_4 \qquad C \supseteq X_3.$$

Since $Hom(X_1,X_3) \subseteq Hom(X_1,X_2) \subseteq P$, and ρ is a representation of P, the groups $\rho(Hom(X_1,X_3))$ and $\rho(Sp(W_4)\cap P)$ commute with one another. Hence we finally need only to verify that $\rho(Hom(X_1,X_3))$ commutes with $\rho(s)$. But this is what we noted above to be true. Hence the claim is proved.

Let P_5 be the parabolic subgroup of Sp(W) consisting of elements which stabilize X_5. We have a decomposition

$$P_5 = M_5N_5 = (GL(X_5)\times Sp(W_4))\cdot N_5 \tag{2.23}$$

analogous to equations (1.10) and (1.11). The fact that $\rho(GL(X_5))$ and $\rho(Sp(W_4))$ commute with one another implies that ρ defines a representation of P_5, as one checks easily using the decomposition (2.23).

If $r \leqslant n-2$, then $\dim X_2 \geqslant 2$, so there is the possibility of a non-trivial decomposition (2.21). Thus we can find a P_5 distinct from P and P_1. So we know ρ defines a representation on at least 3 distinct maximal parabolic subgroups, P, P_1, and P_5, of Sp(W). But this implies that G_8, the group generated by $\rho(P)\cup\rho(P_1)$, must be a homomorphic image of Sp(W), because Sp(W) is the amalgamated free product of P, P_1 and P_5. That is, the group generated by P, P_1 and P_5 subject to relations $y = z$ when $y \in P$, $z \in P_1$ and y and z are in fact the same element of $P\cap P_1$, along with analogous relations for the pairs P, P_5 and P_1, P_5, is exactly Sp(W). That this is so is a consequence of the following result of Tits [T].

Let G be a group with B·N pair. Then G is generated by a (Borel) subgroup B and certain elements s_i. The subgroup generated by B and k of the element s_i is called a standard parabolic subgroup of rank k. The parabolic subgroups generated by B and all but one of the s_i are called standard maximal parabolic subgroups. The result of Tits that we want says G is the amalgamated free product of its standard parabolic subgroups of rank 2. It follows that G is the analgamated free product of any three distinct maximal standard parabolic subgroups. For clearly any rank 2 standard parabolic subgroup must be contained in at least one of three distinct maximal standard parabolics. Hence all the relations defining the amalgamated free product of the rank 2 standard parabolics also hold in the amalgamated product of three maximal parabolic subgroups. Hence this latter product must also be G.

Applying this general result on B-N pairs to our situation tells us that the representation ρ we have constructed is indeed a representation of Sp(W). This concludes the proof of Theorem 1.4.

Remarks: a) The final part of this proof, relying as it does on generators-and-relations arguments, has some aesthetic drawbacks. And it is not very explicit. It might be preferable to write down formulas for the remaining operators in $\rho(Sp(W))$ and verify they define a representation. This approach might also yield the result when dim $X_2 = 1$; however, it would probably be a lot of work.

b) The proof above, as explicit as it is in parts, leaves the full structure of representations of Sp of small rank still somewhat mysterious. In the analogous theory for GL_n, a student of mine, R. Scaramuzzi [Sc], has shown that there is a simple alternative description of representations of small rank: they are all realized as induced representations from appropriate parabolic subgroups. For Sp, this is definitely not so.

3. AN EXAMPLE

We conclude with an example, the case of the trivial representation of the orthogonal group O. In this case the space of functions on which the representation is realized is just $L^2((O\backslash GL(X_1))\times Hom(X_2,X_1))$. We can embed $O\backslash GL(X_1)$ as an open set in $S^{2*}(X_1)$, namely as the $GL(X_1)$-orbit of the inner product β under the natural action. Thus we are dealing with a space of functions on $S^{2*}(X_1)\times Hom(X_2,X_1)$.

Let $\dim X_1=m$ and $\dim X_2=n$. Introduce coordinates u_i on X_1 and v_j on X_2. Then $S^{2*}(X_1)$ can be written as the m×m symmetric matrices. Denote such a matrix by A and let its entries be a_{ij} (with $a_{ij}=a_{ji}$). Similarly $Hom(X_2,X_1)$ becomes the space of m×n matrices T with coordinates t_{ij}. We will compute the action of the Lie algebra sp in these coordinates.

The action of $g\in GL(X_1)$ is given by

$$\rho(g)f(A,T) = |\det g|^{\alpha}f(g^tAg,g^{-1}T)$$

with $\alpha=m+1-n$. Differentiating this action gives the differential operators

$$\sum_k a_{ki} D_{a_{kj}} + a_{ik} D_{a_{jk}} - \sum_{\ell} t_{j\ell} D_{t_{i\ell}} + (m+1-n)\delta_{ij}$$

where $D_{a_{kj}}$ indicated partial differentiation with respect to a_{kj}, and similarly for $D_{a_{jk}}$, $D_{t_{i\ell}}$, etc.

The action of $Hom(X_2,X_1)$ is just by translation and so gives rise to the differential operators

$$D_{t_{i\ell}}$$

(We are using t_{ij} rather than x_{ij} as coordinates on Hom (X_2,X_1).)

The action of $S^2(X_1)$ is just by multiplications. One gets the operators of multiplication by

$$ia_{jk}.$$

The action of $X_1 \otimes X_2$ is likewise by multiplications, by functions

$$i \sum_{\ell} t_{\ell j} a_{\ell k}.$$

And the action of $S^2(X_2)$ is by multiplications by functions

$$i \sum_{k,\ell} t_{kj} a_{k\ell} t_{\ell m}.$$

All these formulas result from differentiating the formulas giving the action of the corresponding piece of Sp(W). If we differentiate formula (2.17) for the operators of $\mathrm{Hom}(X_1, X_2)$, taking into account our new coordinates for $O \times GL(X_1)$, we get the operators

$$mt_{k\ell} + \sum_{ij} a_{ij} t_{j\ell} D_{a_{ik}} + \sum_{ij} a_{ij} t_{j\ell} D_{a_{ki}} - \sum_{ij} t_{i\ell} t_{kj} D_{t_{ij}}.$$

And differentiating formula (2.15) yields the following set of operators coming from $gl(X_2)$.

$$m\delta_{jk} + \sum_{\ell} t_{\ell j} D_{t_{\ell k}}$$

To describe the operators coming from the part of *sp* occupied by C_2 we can not simply differentiate formula (2.20) since this doesn't describe operators near the identity. Instead we adapt to the present situation the formulas of [Sa] describing the oscillator representations. The appropriate operators are

$$i \sum_{k,\ell} (a^{-1})_{k\ell} D_{t_{kj}} D_{t_{\ell m}}$$

where $(a^{-1})_{k\ell}$ indicates the entries of the matrix a^{-1}, the inverse of a.

To describe the operators coming from the remaining parts of *sp*, namely from the elements of the form R_{12} or C_1 in the matrix (2.4), we take commutators. Again refering to (2.4), a commutator of T_{21} and C_2 produces R_{12}, and a commutator of T_{21} and R_{12} produces C_1.

Taking the commutators of operators from T_{21} and C_2 yields operators

$$m\sum_i (a^{-1})_{i\ell} D_{t_{im}} + \sum_i (D_{a_{i\ell}} + D_{a_{\ell i}}) D_{t_{im}} - \sum_{i,j,k} (a^{-1})_{ij} D_{t_{jm}} T_{\ell k} D_{t_{ik}}$$

for R_{12}.

Finally, taking commutators of T_{21} with these operators yields

$$\sum_{ij} a_{ij} (D_{a_{ik}} + D_{a_{ki}})(D_{a_{j\ell}} + D_{a_{\ell j}}) + \sum_{ij} (a^{-1})_{ij} (\sum_\alpha t_{k\alpha} D_{t_{j\alpha}})(\sum_\beta t_{\ell\beta} D_{t_{i\beta}})$$
$$- \sum_i (D_{a_{ik}} + D_{a_{ki}})(\sum_\alpha t_{\ell\alpha} D_{t_{i\alpha}}) - \sum_i (D_{a_{i\ell}} + D_{a_{\ell i}})(\sum_\alpha t_{k\alpha} D_{t_{i\alpha}})$$
$$- (m-1)\sum_i (a^{-1})_{ik} (\sum_\alpha t_{\ell\alpha} D_{t_{i\alpha}}) - m\sum_i (a^{-1})_{i\ell} (\sum_\alpha t_{k\alpha} D_{t_{i\alpha}}) + 3m(D_{a_{k\ell}} + D_{a_{\ell k}}) + m(m-1)(a^{-1}$$

REFERENCES

[H] R. Howe, On a notion of rank for unitary representation of the classical groups, Harmonic Analysis and Group Representations, C.I.M.E. II Ciclo 1980, Cortona-Arezzo, A. Figà-Talamanca coord., Liquori Editore, Naples, 1982, 223-331.

[H2] R. Howe, On the role of the Heisenberg group in harmonic analysis, B.A.M.S. (New Series) v.3(1980), 821-843.

[M1] G. Mackey, Infinite dimensional group representations, B.A.M.S., v.69(1963), 628-686.

[M2] G. Mackey, Unitary representations of group extensions I, Acta Math., v.99(1958), 265-311.

[M3] G. Mackey, Induced representations of locally compact groups II, Ann. of Math., v.58(1953), 193-221.

[M4] G. Mackey, The Theory of Unitary Representations, Chicago Lectures in Mathematics, University of Chicago Press, Chicago and London, 1976.

[Ra] R. Rao, On some explicit formulas in the theory of Weil representation, preprint.

[Ri] M. Rieffel, Induced representations of C^*-algebras, Adv. Math., 18(1979), 176-257.

[Sa] M. Saito, Représentations unitaires des groupes symplectiques, J. Math. Soc. Japan, 24(1972), 232-251.

[Sc] R. Scaramuzzi, Yale Doctoral Dissertation, 1985.

[Sh] D. Shale, Linear symmetries of free boson fields, T.A.M.S., 103(1962), 149–167.

[T] J. Tits, Groups semi-simples isotropes, Colloque sur la théorie des groupes algebriques (Brussels, 1962), Gauthier–Villars, Paris, 1962, 137–147.

[W1] A. Weil, Sur Certains groupes d'opérateurs unitaires, Acta Math. III, (1964), 143–211.

[W2] A. Weil, Sur la formula de Siegel dans la théorie des groupes classiques, Acta Math. 113(1965), 1–87.

DUAL VECTOR SPACES

Irving Kaplansky

Dedicated to George Mackey

1. **INTRODUCTION**. I imagine that most speakers at this Mackeyfest, if they refer to his work, will concentrate on representations of locally compact groups. This subject, of course, constitutes the main body of his research.

If one checks out the entries in the 1940–1959 cumulative index of Mathematical Reviews one finds that his first seven papers concerned linear spaces. Paper number eight, dated 1946, has the title "A remark on locally compact abelian groups" and thereafter linear spaces yield virtually completely to locally compact groups and their representations.

The centerpiece of his linear space era was his thesis. It appeared in two parts [13] and [14]. He and I were graduate students together at Harvard. At the time I had only a general idea of what he was working on; it was only later that I acquired something like a full appreciation.

What Mackey accomplished was an extensive study of a new subject: dual vector spaces (he called them "linear systems"). Independently, and nearly simultaneously. Diedonné launched the same field in [1] and [2].

Forty or so years have now passed. This seems to me to be a perfect occasion for me to review some of the ways I have found dual vector spaces popping up in a variety of contexts. In §§3–7 I exhibit no fewer than five such contexts.

2. **DEFINITIONS, CLOSED SUBSPACES.** In Mackey's dual vector spaces the field of scalars was the field of real numbers. This was natural, since his real objective was the study of locally convex topological vector spaces. However, in the algebraic portion of his work the field is irrelevant. Indeed, it can even be a division ring.

The basic setup is thus as follows: a left vector space V over a division ring D, a right vector space W over D, and a bilinear function (,): V × W ⟶ D. One assumes non-degeneracy: no non-zero element of V(W) annihilates all of W(V). For any subspace S of V or W we form the annihilator S' in the other space. S" is the closure of S, and S is closed is S = S". Priming sets up an anti-isomorphism between the lattices of closed subspaces of V and W.

If S and T are closed subspaces of V, S + T is usually not closed. Mackey proved the equivalence of the following two statements: (a) the lattice of closed subspaces of V is modular, (b) the algebraic union of any two closed subspaces of V is closed. Note that it is equivalent to make these statements about W.

Ornstein [19] studied the modular case. Among his results the following is noteworthy: modularity implies that on one side of the ledger all subspaces of countable dimension are closed and on the other none are closed.

Three purely algebraic examples of dual vector spaces come to mind at once. In the first W is the full dual of V; in the second V is the full dual of W; in the third there are dual bases, i.e., bases v_i, w_i such that (v_i, w_j) is given by the Kronecker delta. Direct sums can be used to construct further examples.

Mackey proved two important results in the case where V and W both have countable dimension. The structure is fully determined by the existence of dual bases; furthermore, any closed subspace is a direct summand (in the obvious sense). This last result was the key to Mackey's generalization [15] of a theorem of Murray [18]: Mackey proved that for any closed subspace of a separable normed linear space there exists a second closed subspace such that the intersection is 0 and the union is dense. Lindenstrauss [12] added the final touch by showing that the assumption of separability cannot be deleted.

3. <u>SIMPLE RINGS WITH MINIMAL IDEALS.</u> Dieudonné [2] discovered a one-to-one correspondence between dual vector spaces and simple rings that possess a minimal one-sided ideal. Given dual vector spaces V and W, the ring in question is the ring of all linear transformations on V that have finite-dimensional range and admit an

adjoint on W. The theory was extended to primitive rings by Jacobson [8, th. 8]; see also [9, p. 75]. A very direct way of extracting the dual vector spaces from the ring appears two pages later in [9]: with e a primitive idempotent in R, the vector spaces are eR and Re and the pairing is simply multiplication.

With suitable countability assumptions the vector spaces have countable dimension and then, by Mackey's theorem, the division ring is the only invariant.

4. __ABELIAN GROUPS.__ The observations in this section arose about ten years ago in conversations with Mark Gordon, a graduate student at Chicago.

Let G be an abelian p-group with no elements of infinite height, P its socle (the set of elements x with $px = 0$). P is a vector space over the field GF(p). In P we have the descending sequence $\{p^iG \cap P\}$ of subspaces with intersection 0.

More generally, in an arbitrary vector space V let there be given a descending sequence $V \supset V_1 \supset V_2 \supset \ldots \supset V_n \supset \ldots$ with $\cap V_i = 0$. There is a natural space W paired with V, consisting of all linear functions on V that vanish on some V_i.

When this is applied to G and P we get dual vector spaces P, Q over GF(p), providing an invariant attached to G.

In continuing the discussion I shall stick to the case where Q has countable dimension. This happens precisely when the Ulm invariants of G are finite, i.e., each $p^{i+1}G \cap P$ has finite codimension in $p^iG \cap P$. In this case, it is a fact that all possible dual vector spaces arise from suitable groups; this can be inferred from the material on page 7 of [5], which, in turn, comes from the paper [7] of Hill and Megibben. Now specialists in abelian group theory have long felt that p-groups with no elements of infinite height exist in such abundance that a structure theory is unthinkable. The same sentiment exists about dual vector spaces; indeed [13, p. 156] we can quote Mackey himself:

"...the problem of classifying completely all linear systems...appears to be hopelessly difficult..."

There is this to be said by way of consolation: if you have two unsolvable problems on your hands, you are making some progress if you find that they are more or less equivalent.

A final comment applies to the case where G itself is countable, so that both P and Q have countable dimension. The theorem of Mackey cited above asserts that P and Q then have dual bases. This makes contact with Prüfer's theorem that a countable p-group with no elements of infinite height is a direct sum of cyclic groups. (Note, however, that we are covering only the case of finite Ulm invariants: Prüfer's theorem is valid without this restriction.)

5. <u>CERTAIN LOCALLY COMPACT ABELIAN GROUPS.</u> I shall reveal at once the force of the adjective "certain": it means that $p^2G = 0$ for some prime p. Over thirty years ago I announced in an abstract [11] that the structure of such a group is determined by a pair of dual vector spaces over GF(p). It is late, but perhaps not hopelessly late to give some details. However, I will leave to the reader the largely routine task of putting additional flesh on the following skeleton.

Let G then be locally compact abelian with $p^2G = 0$. Write G* for the character group.

One knows that G is totally disconnected and therefore possesses a compact open subgroup H. Write J for the annihilator of H in G*; J is compact open in G*. We proceed to detach direct summands.

H is expressible as $H_0 \oplus H_1$, where $pH_1 = 0$ and H_0 is a direct product of copies of $Z/(p^2)$. We claim that H_0 is a direct summand (algebraically and topologically) of G. To see this we expand H_1 to K, maximal with respect to disjointness from H_0. As a matter of pure algebra $H_0 + K$ is all of G. A brief argument, taking advantage of the openness of $H_0 + H_1$, shows that K is closed. In conjunction with the compactness of H_0, this shows that the decomposition $G = H_0 \oplus K$ is topological.

We discard the summand H_0 and change notation. As a result we may assume $pH = 0$. By performing the decomposition relative to G^* we achieve $pJ = 0$. Since J is the character group of G/H we have $p(G/H) = 0$, $pG \subset H$.

The purpose of the next decomposition is to arrange that pG shall be dense in H. Let T be a suitable complement within H of the closure $\overline{pG}$ of pG. We claim that T is a direct summand of all of G. For this purpose, $\overline{pG}$ is enlarged to L, maximal with respect to disjointness from T. That $G = T \oplus L$ algebraically and topologically is argued very much as above. After discarding T and changing notation we have $\overline{pG} = H$, as desired. By working in G^* in the same way we also achieve $\overline{pG^*} = J$.

We are ready to exhibit the dual vector spaces. The vector spaces used are G/H and G^*/J. The pairing is obtained as follows: lift elements of G/H and G^*/J to representatives in G and G^*, take the character pairing, and multiply by p. That this is well defined and yields a non-degenerate pairing follows from the reductions made. The final point (which I omit in toto) is that one can reconstruct the group from the dual vector spaces. In sum: G is a direct sum of a compact group, a discrete group, and a group determined (as just mentioned) by a pair of dual vector spaces.

There is a small disclaimer that has to be made: the pair of dual vector spaces is not quite an invariant of G. Since the compact open subgroup H can be changed (by a finite amount, so to speak) the true invariant is the equivalence class of the dual vector spaces under finite-dimensional changes. It is a classical question, which remains open even for Banach spaces, whether stability holds for such finite-dimensional changes.

When G is separable both of the dual spaces have countable dimension and Mackey's theorem asserts uniqueness. This is in accord with the results of Vilenkin [21]; furthermore, we see that his assumption of separability was indespensable.

6. **LATTICES.** In the lattice approach to finite-dimensional projective geometries the relevant lattices are indecomposable complemented modular lattices of finite length. Ornstein [14] dropped the assumption of finite length and studied complete atomic complemented modular lattices. (An atom is an element directly above the bottom element of the lattice; a lattice is atomic if every element is a union of atoms.) He found that such a lattice is isomorphic to the lattice of closed subspaces attached to a suitable pair of dual vector spaces. It is clear what "suitable" means: we must have the modular law and we must have the property that every closed subspace is a direct summand (Ornstein calls this "splittability"). The case where one space is the full dual of the other is an eligible candidate. There are others. Let α be a cardinal number which is not an inaccessible limit cardinal. Fix a basis $\{v_i\}$ in V and take W to be the set of all linear functions on V which are different from 0 on less than α of the v's. These are apparently all the modular splittable pairs known to this day.

Here is a modest project that I imagine someone will undertake some day: study the lattices corresponding to general pairs of dual vector spaces. A key axiom might well be the modular law for elements of "finite height".

7. **INNER PRODUCT SPACES.** An inner product space is a single vector space, equipped with a bilinear scalar-valued function. The interesting cases are where one assumes symmetry or skew-symmetry.

An inner product space obviously names a pair of dual vector spaces. More significant is the fact that one can proceed in the reverse direction. Given a pair V, W we use the vector space $V \oplus W$ and, in the symmetric case, the inner product

$$[v + w, v' + w'] = (v, w') + (v', w).$$

This construction is used in Savage's paper [20] to exhibit, over the real numbers, an inner product space which cannot be expressed as a direct sum of two subspaces, one positive definite and the other negative definite (the construction is attributed to Mackey).

Here is a theorem that merits a comment: only in the finite-dimensional case is the lattice of closed subspaces of an inner product space modular. This is noted on page 40 of [6], credited to Keller (unpublished). But it is an immediate consequence of the theorem of Ornstein (cited above) asserting that in a modular pair of dual vector spaces the countable-dimensional subspaces are closed in precisely one of the two spaces.

If we change the formula above to

$$[v + w, v' + w'] = (v, w') - (v', w)$$

we get a skew-symmetric inner product space. It is standard that in the finite-dimensional case all skew-symmetric forms arise in this way, and by Mackey's methods it is easy to extend this to the case of countable dimension. At the conference Irving Segal asked me whether countability can be dropped.

The answer is "no". I have a recollection that Donald Ornstein constructed a counter-example way back in the 1950's. He shares this recollection, but neither of us seems to have a written record of it. Perhaps his example was the same as the following.

I shall actually do a little more by constructing a skew-symmetric inner product space A of uncountable dimension in which every totally isotropic subspace has countable dimension. This does the job, for if $A = V \oplus W$ as above, then at least one of V and W has uncountable dimension and is totally isotropic.

To any field k adjoin an uncountable number of indeterminates ($\aleph_1$, if you like to be precise); the resulting field F is the base field to be used. Erect a vector space A over F with basis $\{u_i\}$ having the same cardinal number as the number of indeterminates. Linearly order the index set over which i ranges. Set up a one-to-one correspondence between the indeterminates and the pairs i,j with $i<j$; write x_{ij} for the corresponding indeterminate. Define an inner product by

$$(u_i, u_i) = 0$$

$$(u_i, u_j) = -(u_j, u_i) = x_{i,j} \text{ for } i<j.$$

This is the proposed inner product space.

Suppose that on the contrary A contains S, an uncountable linearly independent set of elements, any two orthogonal. Take any countable subset T of S. We shall exhibit a member of S not orthogonal to T.

Any member of S is a finite linear combination of u's with coefficients rational functions of the x's. By multiplying by the product of the denominators, we can assume that the coefficients are polynomials in the x's. To each such member of T we associate a finite number of u's as follows: first, those which occur in its expression, and then in addition, u_i and u_j for each indeterminate $x_{i,j}$ which actually occurs in the polynomials serving as coefficients. When we collect all these for all the members of T we get a certain countable set U of u's. There must exist an element $s \in S$ involving a u_r which is not in U (this is because S is uncountable). The polynomial coefficients of s involve a finite number of indeterminates and to each of these a pair of u's corresponds. We can pick an element $t \in T$ involving a u_p other than these and also distinct from the u's occuring in s itself. Now we contemplate the statement $(s,t) = 0$. From the contribution (u_r, u_p) the indeterminate x_{rp} arises. Nowhere else in the computation of (s,t) will x_{rp} occur. This is impossible, for we would have a polynomial equation of the form $x_{rp}f_1 + f_2 = 0$, where f_1 and f_2 are polynomials in <u>other</u> indeterminates.

An analogous discussion in the symmetric case is probably feasible, but I shall not undertake it here. One should note that one would have trivial examples not representable in the $V \oplus W$ fashion by having no isotropic vectors at all; of course, this remark does not apply if the base field is algebraically closed.

8. <u>DUAL MODULES.</u> In this section I say a few brief words about the generalization to dual modules over rings. In [10] I took the first steps in extending Mackey's theory to dual modules over a DVR (discrete valuation ring). Somewhere in the future I foresee further generalizations. I recommend to the reader a look at the case of complete local rings; Matlis's duality [17] should be quite helpful.

9. **TWO BIBLIOGRAPHIC NOTES.** (a) Dual vector spaces are useful in the study of non-unitary representations of locally compact groups; see pages 657-659 of [16].

(b) In two recent papers ([3], [4]) large parts of the Mackey theory are used in a general study of differentiation. (I am indebted to Saunders Mac Lane for these references.)

REFERENCES

[1] J. Dieudonné, La dualité dans les espaces vectoriels topologiques, Ann. Sci. École Norm. Sup. **59** (1942), 107–139.

[2] ___________, Sur le socle d'un anneau et les anneaux simples infinis, Bull, Soc. Math, France, **70** (1942), 46–75.

[3] A. Frölicher, Smooth structures, pp. 69–81 in Category Theory, Springer Lecture Notes, no. 962, 1982.

[4] A. Frölicher, B. Gisin, and A. Kriegl, General differentiation theory, preprint, 28 pp.

[5] L. Fuchs, Infinite Abelian Groups, vol. II, Academic Press, 1973.

[6]. H. Gross, Quadratic Forms in Infinite Dimensional Vector Spaces, Birkhäuser, 1979.

[7] P. Hill and C. Megibben, On primary groups with countable basic subgroups, Trans. Amer. Math. Soc., **124** (1966), 49–59.

[8] N. Jacobson, On the theory of primitive rings, Ann. of Math. **48** (1947), 8–21.

[9] ___________, Structure of Rings, Amer. Math. Soc. Coll. Publ. vol. 37, revised edition, 1964.

[10] I. Kaplansky, Dual modules over a valuation ring, I, Proc. Amer. Math. Soc. **14** (1953), 213–219.

[11] ___________, ibid II, abstract, Bull, Amer. Math. Soc. **59** (1953), 154.

[12] J. Lindenstrauss, On subspaces of Banach spaces without quasicomplements, Israel J. of Math. **6** (1968), 36–38.

[13] G.W. Mackey, On infinite-dimensional linear spaces, Trans. Amer. Math. Soc. **57** (1945), 155–207.

[14] ___________, On convex topological linear spaces, Trans. Amer. Math. Soc. **60** (1946), 519–537.

[15] ___________, Note on a theorem of Murray, Bull. Amer. Math. Soc. **52** (1946), 322–325.

[16] ___________, Infinite-dimensional group representations, Bull. Amer. Math. Soc. **69** (1963), 628–686.

[17] E. Matlis, Injective modules over Noetherian rings, Pac. J. of Math. **8** (1958), 511–528.

[18] F.J. Murray, Quasi-complements and closed projections in reflexive Banach spaces, Trans. Amer. Math. Soc. **58** (1945), 77–95.

[19] D. Ornstein, Dual vector spaces, Ann. of Math. Soc. **69** (1959). 520–534.

[20] L.J. Savage, The application of vectorial methods to metric geometry, Duke Math. J. **13** (1946), 521–528.

[21] N. Vilenkin, Direct decompositions of topological groups, I, Mat. Sbornik **19** (1946), 85–154; Amer. Math. Soc., Translation no. 23, 1950.

EXPONENTIAL DECAY OF CORRELATION COEFFICIENTS FOR GEODESIC FLOWS

by Calvin C. Moore[(1)]

Dedicated to George W. Mackey

Abstract

We obtain exponential decay bounds for correlation coefficients of geodesic flows on surfaces of constant negative curvature (and for all Riemannian symmetric spaces of rank one), answering a question posed by Marina Ratner. The square integrable functions on the unit sphere bundle of M are allowed to satisfy weak differentiability conditions. The methods are those of unitary representation theory and invoke the notion of Sobelev vectors of representations. In the course of the discussion we obtain a new characterization of tempered irreducible representation of semi-simple groups.

---

(1) Supported in part by NSF Grant DMS-83-08252 and DMS-8120790.

1980 Mathematics Subject Classification 22E46, 58F17.

§1.

Marina Ratner [19] raised the following question concerning geodesic flows: suppose that M is a Riemann surface of constant negative curvature equal to -1 (compact or finite volume) and that g_t, $t \in \mathbb{R}$ is the geodesic flow on the unit circle bundle SM over M. Suppose further that φ and ψ are two square integrable functions on SM, both perpendicular to the constants, and consider the inner product

$$\int \varphi(g_t x)\bar{\psi}(x)dx = (g_t{}^*\varphi,\psi).$$

Then as $|t| \longrightarrow \infty$, it is well-established that this tends to zero, but in addition Ratner wanted to know that under some mild continuity conditions on φ and ψ (say Hölder of some exponent) there was an exponential bound on the decay

$$|(g_t{}^*\varphi,\psi)| \leqslant Ce^{-k|t|}.$$

For very smooth functions φ and ψ, say of class C^∞, one approaches the question using the classical translation [9] of the problem into a problem in group representation and applying known exponential bounds on the decay of matrix coefficients of smooth vectors of representations of semi-simple Lie groups [10] (cf. also [24], Chapter 9), [16], [22], [23], and [3]. In the case of interest $G = SL_2(\mathbb{R})$, results in [1] would suffice.

Our goal here is to present a very simple and elementary argument that establishes the desired exponential decay whenever φ and ψ satisfy Hölder conditions of exponent greater than 1/2; in fact these conditions only need be satisfied in the "fiber direction" in SM. Specifically, we view SM as a principal circle bundle over M and we demand for $x \in SM$ that

$$|\varphi(t\cdot x) - \varphi(x)| \leqslant K(x)|t|^{\alpha}$$

where t is in the circle group, $t\cdot x$, the action of t on $x \in SM$ and $|t|$ the distance from t to 1 in the circle. We do not need that $K(x)$ be bounded, only that $|K|^2 = \int |K(x)|^2 dx$ be finite. The result is as follows.

Theorem 1.1. Let φ and ψ be in $L^2(SM)$, orthogonal to the constants, and satisfying Hölder condition in the fiber direction described above for some $\alpha > 1/2$. Then if g_t is the geodesic flow

$$|(g_t{}^*\varphi,\psi)| \leqslant Le^{-c|t|}$$

where c depends only on M and L depends on the Hölder norms of φ and ψ.

We shall see that the constant c is essentially determined by the lower bound of the spectrum of the Laplacian on M. In addition one has analogous power estimates for decay of correlation coefficients for the horocycle flow.

Theorem 1.2. If h_t is the horocycle flow on SM and if φ and ψ are as in Theorem 1, then

$$|(h_t{}^*\varphi,\psi)| \leqslant L'|t|^{-c}, \quad t \text{ large}.$$

These results flow from results about unitary representations, valid not just for SL_2, but for all semi-simple groups. Indeed the same techniques can be used to obtain asymptotic decay theorems for quite general flows on homogeneous spaces G/Γ, cf. [4]. However we shall not go into these extensions here in detail.

The proofs involve introduction of a kind of Hölder and Sobelev spaces of vectors in a unitary representation space which have properties of fractional differentiability in the direction of the maximal compact subgroup. We use this together with a characterization of tempered representations to obtain the decay estimates for this class of representations. Our characterizations of tempered representations (Proposition 3.3) may be of some independent interest. One may then reduce the asymptotic estimate in general to the tempered case via a tensor power argument and results of Cowling [5].

We remark that after these results were presented at the conference, Roger Howe showed us a very interesting and elegant alternative method of obtaining such estimates based in part on ideas of Cowling and Herz. This is rather disjoint from our presentation.

§2.

Suppose that K is a compact Lie group. We fix a two sided invariant Riemannian metric on K; $|k|$ will denote the distance from $k \in K$ to the identity element and Δ will denote the Laplace operator of the metric. We will think of Δ as an element of the universal enveloping algebra of K.

Now let π be a unitary representation of K on a Hilbert space $H(\pi)$.

Definition 2.1. A vector v in $H(\pi)$ is a Hölder vector of index α, $0 < \alpha < 1$, if $|\pi(k)v-v| \leqslant A|k|^{\alpha}$ for all $k \in K$.

To cover exponents larger than one we shift to closely related concept.

Definition 2.2. A vector v in $H(\pi)$ is a Sobelev vector of index α if v is in the domain of the operator $(1-d\pi(\Delta))^{\alpha/2}$ where $d\pi$ is the corresponding infinitesimal representation of the universal enveloping algebra.

For $0 < \alpha < 1$ the relation between these notions is well known, [20], Proposition 4, p.139.

Proposition 2.3. If t is a Hölder vector of index α, $0 < \alpha < 1$, then t is a Sobelev vector of index α' for any α', $0 < \alpha' < \alpha$.

We record some elementary consequences of the definition.

Proposition 2.4. If the representation π is a direct integral of representations π^x on Hilbert spaces $H(\pi^*)$, and if $v \in H(\pi)$ has a decomposition $v = \int v^x du(x)$, then if v^x is a Hölder vector of exponent α and Hölder constant K(x), v is a Hölder vector if K(x) is square integrable. If v^x is a Sobelev vector of index α, then v is if $|1-d\pi^x(\Delta)v^x|$ is square integrable.

The simple proof is omitted.

If for instance K acts on a manifold Y preserving a measure ν and we form $H(\pi) = L^2(X)$ with π given by the induced action, this representation can be decomposed as a direct integral over the orbit space $K\backslash Y = X$ with the Hilbert spaces $H(\pi^*)$ being L^2 of the orbit.

Corollary 2.5. Let φ be in $L^2(Y)$ and suppose $|\varphi(k\cdot y)-\varphi(y)| \leqslant L(y)|k|^\alpha$. Then if L is a square integrable function on Y, φ is a Hölder vector.

The following will be crucial for us.

Proposition 2.6. Suppose that $H(\pi) = L^2(K,H_0)$ the space of square integrable H_0 valued functions on K with π being given by left translation. If v is a Sobelev vector of index $\alpha > (\dim K)/2$, then v as an element of $L^2(K,H_0)$ is represented by a (weakly) continuous function from K to H_0. In particular if dim K = 1, and if v is a Hölder vector if index $\alpha > 1/2$, v is represented by a (weakly) continuous function. Moreover, the supremum or L^∞ norm of the function v can be estimated in terms of the ordinary L^2 norm and the Hölder or Sobelev norm of v.

Proof. Except for the functions being vector valued, this is a standard Sobelev space fact, [20, p.124]. The extension to vector valued functions is clear by taking inner products. The last statement follows from the first plus Proposition 2.3. The estimates on the L^∞ norm are also standard by use of the closed graph theorem.

§3.

The second ingredient in the proof of our results is the notion of a tempered representation.

Definition 3.1. A unitary representation of π of a locally compact group is tempered if π is weakly contained, in the sense of Fell [8], in the regular representation of G.

This means specifically that any matrix coefficient of π may be approximated uniformly on compact sets of G by convex combinations of matrix coefficients of the regular representation. We remind the reader of the well known fact that every representation of G is tempered if and only if G is amenable (cf. [7]). Tempered irreducible representations of semi-simple groups were studied at great length by Harish Chandra and others and play a key role in representation theory for these groups. Our goal here is a characterization of tempered representations in terms of being subrepresentations of other representations.

Recall that in [18], we classified the maximal amenable subgroups of a semi-simple group. Suppose that G has finite center and real rank r. Then any connected amenable subgroup is contained in a maximal connected subgroup and these in turn fall into exactly 2^r conjugacy classes. Any amenable subgroup is contained in a maximal connected one up to a group of finite index. The classification is as follows and for this we assume G is algebraic -- the modifications needed in general are clear. One fixes a Borel subgroup B = MAN with its usual decomposition. For any parabolic subgroup $P \supset B$ (of which there are 2^r, r = dim A), write its Langlands decomposition $P = M_P A_P N_P$ where $A_P \subset A$, $N_P \subset N$, and M_P is reductive with compact center. Now let K_P be the maximal compact subgroup of M_P. It is evident that $S_P = K_P A_P N_P$ is amenable, being a compact extension of a solvable group. It is also a maximal amenable subgroup and its connected compact S_P^0 is a maximal connected amenable subgroup (Theorem 3.4 of [18]). Notice that at the extremes, P = B or P = G, the corresponding maximal amenable subgroups are the Borel subgroup and the maximal compact subgroup respectively.

The following characterizes irreducible tempered representations.

<u>**Proposition 3.2.**</u> An irreducible representation π of G is tempered if and only if it is a subrepresentation of a representation of G induced from a unitary finite dimensional representation of one of the 2^r maximal amenable connected subgroup of G described above.

Proof. By continuity in the Fell topologies of the induction process, [8], it follows that inducing a tempered representation from a subgroup yields a tempered representation of the whole group. Since all representation of an amenable group are tempered and since a subrepresentation of a tempered representation is clearly tempered, the "if" part of the Proposition is established.

For the converse, we use the description of irreducible tempered representations π_0 provided by [14], [16], [21] (cf. also [3], IV 3.7) that any such π_0 is a subrepresentation of a representation of G induced from $P = M_P A_P N_P$ by a unitary representation of the form $\pi_0(m \cdot a \cdot n) = \pi_1(m) X(a)$ where X is a character if A_P, and π_1 is an irreducible square integrable representation of M_P. But now π_1 being sqaure integrable is a subrepresentation of the induced by some finite dimensional representation of the maximal compact subgroup K_P. The result now follows.

As a consequence, we have the following fact.

Proposition 3.3. A unitary representation π of G is tempered if and only if π is a subrepresentation of a representation induced by a unitary representation π_0 from the Borel subgroup B.

Proof. The same reasoning used in the first part of the proof of 3.2 shows that any representation of the prescribed form is tempered. To show the converse it suffices to consider an irreducible tempered representation since the general case follows from this one by a direct integral argument. By 3.2, any irreducible tempered π is a subrepresentation of a representation induced from a representation π_0 of $S_P^0 = K_P^0 A_P^0 \cdot N_P$. But π_0 is contained in the result of inducing the restriction π_1 of π_0 to $A_P^0 N_P$ back up to S_P^0. But then $A_P^0 N_P$ is contained in B and so by transitivity of induction, π_1 induced up to B, and the result then induced up to G will contain the original representation π of G as a subrepresentation, and we are done.

Finally, we discuss for a moment a connection between all representations and tempered representations. We first recall a well known fact, cf. [6].

Lemma 3.4. If π is a unitary representation of a locally compact group G on a Hilbert space $H(\pi)$ and if there is a dense subspace D of $H(\pi)$ such that all matrix coefficients $(\pi^{-1}(g)u,v)$, $u,v \in D$, are square integrable on G (left Haar measure), then π is unitarily equivalent to a subrepresentation of $\infty\pi_{reg}$, where π_{reg} is the regular representation. In particular π is tempered.

The following is a formulation of results of Cowling, [5].

Theorem 3.5. Let π be an irreducible representation of a semi-simple group with discrete kernel. Then there is an integer k such that $\pi^{\otimes k}$ (the kth tensor power) is tempered. If G has no simple factor locally isomorphic to SO(n,1) or SU(n,1) then the k can be taken to be the same for all π. Moreover, for any k unitary representations of G non of which contains the trivial representation when restricted to any simple factor of G, $\pi_1 \otimes \pi_2 \otimes \ldots \otimes \pi_k$ is tempered.

Proof. This follows from 2.4.2, 2.5.2 and 2.5.3 of [5] which establish L^p exists for matrix coefficients of such representations. The construction of the D in the tensor product to meet the conditions of Lemma 3.4 if clear.

It is well known that SO(n,1) and SU(n,1) have representations where the k needed above can be arbitrarily large. The spherical complementary series π_s, $0 < s < 1$ for these groups have matrix coefficients that are not in $L^{2/(1-s)}$ and which require a k larger than $(1-s)^{-1}$ before $\pi^{\otimes k}$ is tempered.

For SO(2,1) and a few other low dimensional cases, these are the only non tempered irreducible representations. Hence if π is any representation not containing the trivial representation, then $\pi^{\otimes k}$ is tempered if and only if there is s_0 such that the direct integral decomposition of π does not contain any spherical complementary series for $s > s_0$. Indeed this holds more generally.

Proposition 3.6. For G = SO(n,1) or SU(n,1), and any unitary representation π, $\pi^{\otimes k}$ is tempered for some k if and only if the direct integral decomposition of π does not contain any spherical complementary series π_s for s greater than some s_0.

Proof. This follows from the classification of unitary representations of these groups [11], [12], [13] and the theory of complementary series [14], [16]. The only non tempered representations are complementary series and ends-of-complementary series, [16], p.152. These complementary series with parameter z exist up to a critical index z_c [14], and if one normalizes the critical index of the spherical complementary series to be 1, then critical indices for all other complementary series for fixed G are bounded away from 1, because the integrality conditions in [14] bound z_c from 1 unless it is equal to 1. But for any spherical complementary series, the critical index cannot be equal to 1, otherwise the end-of-complementary series representation π_c would fail to have the property that $\pi_c^{\otimes k}$ is tempered for some k and hence would be the trivial representation, and that would imply that the complementary series had a K-fixed vector and is hence the spherical complementary series.

§4.

With these preliminaries we turn to the main topic, which is to obtain exponential decay estimates for matrix coefficients $(g_t^* u,v)$ where u and v are Sobelev vectors. First we suppose that π is a tempered representation of a semi-simple group G = KAN with Borel subgroup B = MAN. Suppose that v is a Sobelev vector of K of index $\alpha > (\dim K)/2$ in $H(\pi)$. A Sobelev or Hölder vector for K will mean a vector as defined in Section 2 for the representation π restricted to K. By Proposition 2.3, π is a subrepresentation of a representation π_1 induced from a unitary representation λ_1 of B. Then v is a Sobelev vector for K in $H(\pi_1)$. But since $G = K \cdot B$, $H(\pi_1)$ can be viewed as a space of $H(\lambda_1)$ valued square integrable functions on K and this can be arranged so that K as a subgroup of G acts by left translations on this space of functions. Then v in this presentation of $H(\pi)$ as a subspace of $L^2(K,H(\lambda))$ is by Proposition 2.6

a continuous function on K. In particular, it is bounded function and that is all we need to know about v for the result of the argument.

We consider two Sobelev vectors v and w for the action of W and we want to estimate the matrix coefficient $(\pi(g)u,v)$. As usual, we restrict to the case when $g \in A$, and for simplicity we will henceforth take G to be rank 1. Let $\mathfrak{A}$ be the Lie algebra of A, and $\mathfrak{n}$ be the Lie algebra of N. Then $\mathfrak{n} = \mathfrak{n}_1 + \mathfrak{n}_2$ where $[X,n_i] = i\alpha(X)n_i$ for $n_i \in \mathfrak{n}_i$. We fix X so $\alpha(X) = 2$ and let $a_t = \exp(tX)$. This arranges matters so that for $SL_2(\mathbb{R})$, $a_t = \begin{bmatrix} e^t & 0 \\ 0 & e^{-t} \end{bmatrix}$.

Now we examine how $\pi(a_t)$ acts in $L^2(K,H(\lambda))$. This action comes from a spatial action of the one parameter group a_t on the space K viewed as G/AN, together with a cocycle. The cocycle is not unitary as it has to correct for the failure of the spatial action to be measure preserving. The spatial action is best analyzed on the quotient K/M = G/MAN = G/B where M centralizes A. In G/B, A has two fixed points and all other orbits flow toward one fixed point as $t \longrightarrow \infty$ and toward the other as $t \longrightarrow -\infty$. Think of G/B as a sphere with the two fixed points the north and south pole and with a_t flowing along great circles. Using a Bruhat cell in G/B centered at a fixed point, we can determine how fast $a_t \cdot u$ approaches the fixed point in terms of the restricted roots relative to $\mathfrak{A}$.

If L is a cross section for the orbits of A on K (an equator in the picture above), we can represent $L^2(K,H(\lambda))$ as $L^2(A \times L,H(\lambda))$ with A acting by translation on itself plus a cocycle. The K-invariant measure in these coordinates looks like $f(t,\ell)\, dt d\mu_L(\ell)$ where μ_L is a finite measure on L. From the structure above it is clear that $f(t,\ell)$ will decay as $|t| \longrightarrow \infty$ and we can bound the rate of decay knowing how fast $a_t \cdot u$ approaches the fixed points.

An easy classical calculation for $G = SL_2(\mathbb{R})$ shows that $f(t,\ell) = (\cosh(2t))^{-1}$ and in general an easy calculation gives an estimate

$$f(t,\ell) \leqslant c(\cosh(2pt))^{-1}$$

where $\rho(X) = p$ and ρ is half the sum of the positive restricted roots counting multiplicities.

Therefore if we make a unitary transformation from $L^2(K,H(\lambda))$ over to $L^2(A\times L,H(\lambda))$ with measure $dtd\mu_L$, a function $v(k)$ goes over into $\tilde{v}(t,\ell) = v(t,\ell)f(t,\ell)^{1/2}$. Moreover the non unitary part of the cocycle describing the action of A disappears since A now preserves the measure, and the cocycle is a coboundary because A is acting on itself. Once this change is made, the representation π on A looks like

$$(\pi(a_t)\tilde{v},\tilde{w}) = \int_L\int_{\mathbb{R}}(\tilde{v}(t+s,\ell),\tilde{w}(x,\ell))dtd\mu_L(\ell).$$

Now starting with two Sobelev vectors v and w, which we know to be bounded functions on K, we see that the corresponding functions $\tilde{v}$ and $\tilde{w}$ are bounded by constants times $\exp(-2p|t|)$ where the constants depend on the L^2 norms of v and w and their Sobelev norms. Consequently,

$$|(\pi(a_t)v,w)| = |(\pi(a_t\tilde{v},\tilde{w})|$$
$$\leqslant c\int\exp(-2p|t+s|)\exp(-2p|s|)dx$$
$$\leqslant C'\exp(-p|s|)|s|.$$

Therefore, we have proved the following.

<u>**Theorem**</u> <u>**4.1.**</u> Let π be a tempered representation of the rank one group G = KAN, and let $a_t \in A$ and $k \in \mathbb{R}$ as above. Then if u and v are two Sobelev vectors for K of index strictly larger than (dim K)/2, we have bounds

$$|(\pi(a_t)v,w)| \leqslant C_\epsilon \exp(-(p+\epsilon)|t|)$$

for every ϵ where C_ϵ depends on the norms and Sobelev norms of v and w. For $SL_2(\mathbb{R})$, $p = 1$.

For non tempered representations we then obtain the following.

Theorem 4.2. With the notation of 4.1, we let π be a unitary representation such that $\pi^{\otimes k}$ is tempered. Then for Sobelev vectors v and w for K of index strictly greater than (dim K)/2 we have bounds

$$|(\pi(a_t)v,w)| \leqslant C_\epsilon \exp(-((p/k)+\epsilon)|t|)$$

for every ϵ where C_ϵ depends on the norms and Sobelev norms of v and w. Again for $SL_2(\mathbb{R})$, $p = 1$.

Proof. We observe that $v^{\otimes k}$ and $w^{\otimes k}$ are Sobelev vectors of the same index as v and w; to see this, look at the full tensor product with $K\times\ldots\times K$ acting on $H(\pi)^{\otimes k}$. The result following immediately.

Furthermore, we obtain immediately analogous results for decay along nilpotent groups, that is in the N direction. Let $n \in N$ and let $\varphi(n)$ be the unique positive number so that $n = ka_{\varphi(n)}k'$ with k and k' in K. Since the action of K preserves K-Sobelev vectors and does not change the Sobelev norms, we obtain the following.

Theorem 4.3. With notation as in Theorem 4.2, we have bounds for $n \in N$

$$|(\pi(n)v,w)| \leqslant C_\epsilon \exp(-((p/k)+\epsilon)\varphi(n))$$

for pairs of Sobelev vectors of index strictly greater than (dim K)/2.

Since for $n(u) = \exp(yu)$, $y \in \mathfrak{n}$ $\varphi(n(u))$ is for large $|u|$

asumptotic to $c\log|u|$ for some constant, Theorem 4.3 yields power bounds for large $|u|$

$$|(\pi(n(u))v,w)| \leqslant C_\epsilon |u|^{d+\epsilon}$$

for appropriate d. For SL_2, there is a unique one parameter group in N, and it is evident that $d = 1$.

§5.

We return now to proof of the theorems stated in the introduction. We translate Theorems 1.1 and 1.2 into representation theory via the classical work of Gelfand and Fomin [9]. Specifically if $\Gamma \subset SL_2(\mathbb{R})$ is arranged so that $\Gamma/(\pm 1)$ is the fundamental group of the Riemann surface M, then $M = K\backslash SL_2(\mathbb{R})/\Gamma$, and SM the unit circle bundle is $SL_2(\mathbb{R})/\Gamma$. Moreover, the geodesic and horocycle flows on SM are given by actions of the one parameter groups $a(t) = \begin{pmatrix} t & 0 \\ 0 & t^{-1} \end{pmatrix}$ and $n(u) = \begin{pmatrix} 1 & u \\ 0 & 1 \end{pmatrix}$ restrictively.

We want therefore to consider the unitary representation π of $SL_2(\mathbb{R})$ on the space $L^2(SL_2(\mathbb{R})/\Gamma)\ominus(1)$ of functions orthogonal to the constants. What we need to know is that $\pi^{\otimes k}$ is tempered for some k, which by our remarks in Section 3 is the same as π not containing in its decomposition any spherical complementary series representations π_s for s greater than some fixed s_0. However, appearance of π_s discretely in the representation π produces via its K-fixed vectors, functions on $K\backslash SL_2(\mathbb{R})/\Gamma = M$ which will be eigenfunctions of $d\pi(\Delta)$ of eigenvalue $(1-s^2)/4$, where Δ is the second order central element in the enveloping algebra, so that $d\pi(\Delta)$ dropped down onto $K\backslash SL_2(\mathbb{R})/\Gamma = M$ is the Laplacian of M.

Appearance of a direct integral of π_s's in π similarly produces continuous spectrum of $d\pi(\Delta)$. Hence the absence of spherical complementary series with values of s near 1 in π is exactly the same as the absence of spectrum of the Laplacian in a small interval above zero. However, for compact surfaces M, the Laplacian has discrete spectrum tending to infinity so there is a small gap to the first eigenvalue and so $\pi^{\otimes k}$ is tempered for some k. For M of finite volume, Borel and Garland [2] show that the spectrum of the Laplacian also has a gap above zero, and the same argument applies. This proves Theorem 1.1. Theorem 1.2 now follows immediately.

One can obtain an estimate on the k needed so that $\pi^{\otimes k}$ is tempered. Indeed if λ_1 is the lower bound of the spectrum of the Laplacian of M on $L^2(M)\ominus 1$, then k is approximately $2\lambda_1^{-1}$. Thus we can state the following.

<u>**Proposition**</u> <u>5.1.</u> In theorems 1.1 and 1.2, the constant c in the estimates.

$$|(g_t*\varphi,\psi)| \leqslant Le^{-ct}$$

and

$$|(h_t*\varphi,\psi)| \leqslant L'|t|^{-c}$$

can be taken to be approximately $\lambda_1(M)/2$ where $\lambda_1(M)$ if the lower bound of the spectrum of the Laplacian on $L^2(M)\ominus 1$. This means specifically that as the Riemann surface M, and if c(M) is the upper bound of the constants c that work in these estimates, then $2c(M)/\lambda_1(M) \longrightarrow 1$ as $\lambda_1(M)$ tends to zero.

Proof. This follows easily from what we have done and the fact that there are coefficients of the spherical complementary series that decay as slowly as the estimates permit.

Finally, it is evident that the techniques can be used to

investigate decay of coefficients of geodesic flow on other finite volume rank one symmetric spaces $K\backslash G/\Gamma$. The situation is particularly simple when the group is not locally isomorphic to SO(n,1) or SU(n,1) for then there is a fixed k, independent of the subgroup Γ, such that $\pi^{\otimes k}$ is tempered where π as usual is the representation of G on $L^2(G/\Gamma)\ominus(1)$. Hence one can obtain exponential bounds on decay of the coefficients that are independent of which Γ one chooses. For G isomorphic to SO(n,1) or SU(n,1), i.e., for real and complex hyperbolic spaces of higher dimension, we know that for any π, $\pi^{\otimes k}$ tempered for some k is equivalent to the absence of spherical complementary series π_s, $s > s_0$ in the decomposition of π_1 and in turn for π the representation on $L^2(G/\Gamma)\ominus 1$, this is the same, just as for SO(2,1) as having a small gap above 0 in the spectrum of the Laplacian on $M = K\backslash G/\Gamma$. For compact M one has this gap automatically and Borel-Garland [2] provides the gap in the finite volume case.

REFERENCES

[1] V. Bargmann, Irreducible unitary representations of the Lorentz group, Ann. of Math. 48, (1947), 568-640.

[2] A. Borel and H. Garland, Laplacian and the discrete spectrum of an arithmetic group, Amer. J. of Math. v.105, (1983), 309-335.

[3] A. Borel and N. Wallach, *Continuous Cohomology, Discrete Subgroups, and Representations of Reductive Groups*, Ann. of Math. Studies 94, (1980), Princeton University Press.

[4] J. Brezin and C.C. Moore, Flows on homogeneous spaces: a new look, Amer. J. of Math. v.103, (1981), 571-613.

[5] M. Cowling, Sur les coefficients des représentations unitaires des groupes de Lie simple, Lecture Notes in Mathematics 739, (1979), 132-178, Springer-Verlag.

[6] J. Dixmier, *Les C*-Algebres et leur* Représentations, Gautheir-villars, Paris, 1969.

[7] P. Eymard, *Moyennes Invariantes et Représentations Unitaires*, Lecture Notes in Math. 300, (1972), Springer-Verlag.

[8] J.M.G. Fell, Weak containment and induced representation of groups, Canadian Math. J. 14, (1962), 237-268.

[9] I.M. Gelfand and S. Fomin, Geodesic flows on manifolds of constant negative curvature, Uspekhi Matem. Nauk 7, (1952), 118-137.

[10] Harish Chandra, Spherical functions on a semi-simple group, I, Amer. J. of Math. 80, (1958), 241–310.

[11] T. Hirai, On irreducible representations of the Lorentz group of nth order, Proc. Japan Acad. 38, (1962), 83–87.

[12] H. Kraljevic, Representations of the universal covering group of the group SU(n,1), Glasnik Mat., Ser. 3, 8, (28), (1973), 23–72.

[13] ———, On representations of the group SU(n,1), Trans. Amer. Math. Soc. 221, (1976), 433–448.

[14] A.W. Knapp and E.M. Stein, Intertwining operators for semi-simple groups, Ann. of Math. 93, (1971), 489–578.

[15] A.W. Knapp and G. Zuckerman, Classification of irreducible tempered representations of semi-simple Lie groups, Proc. Nat. Acad. Sci., USA 73, (1976), 2178–2180.

[16] ——— and ———, Classification theorems for representations of semi-simple Lie groups, Lecture Notes in Math. 587, (1977), 138–159, Springer-Verlag.

[17] R. Langlands, On the classification of irreducible representations of real algebraic groups, notes, Institute for Advanced Study, Princeton, New Jersey, 1973.

[18] C.C. Moore, Amenable subgroups of semi-simple groups and proximal flows, Israel J. Math. 34, (1979), 121–138.

[19] M. Ratner, Rigidity of time changes for horocycle flows, to appear, Acta Math.

[20] E.M. Stein, *Singular Integrals and Differentiability Properties of Functions*,

Princeton University Press, Princeton, New Jersey, 1970.

[21] P. Trombi, The tempered spectrum of a real semi-simple Lie group, Amer. J. of Math. 99, (1977), 57-75.

[22] P. Trombi and V. Varadarajan, Spherical transforms on semi-simple Lie groups, Ann. of Math. 94, (1971), 246-303.

[23] ——— and ———, Asymptotic behavior of eigenfunctions on a semi-simple Lie group: the discrete spectrum. Acta Math. 129, (1972), 237-280.

[24] G. Warner, *Harmonic Analysis on Semi-simple Lie groups*, II, Grund. Math. Wiss. 189, Springer, 1972.

Mathematical Sciences Research Institute
and
Department of Mathematics, UC Berkeley

Berkeley, California 94720

LATTICES IN U(n,1)

G. D. Mostow

To George Mackey

1. INTRODUCTION

Since this paper is related to Mackey's influence in a roundabout way, a few words are appropriate to explain the link: ergodic theory.

G. A. Margulis' remarkable proof of the arithmeticity of irreducible lattices in semi-simple groups of $\mathbb{R}$-rank > 1 hinged on replacing my geometric proof of strong rigidity by an ergodic theoretic proof of a property much stronger in the case $\mathbb{R}$-rank > 1, which I named "super-rigidity." At that time, non-arithmetic lattices in PO(n,1) were known, at least for $n \leqslant 5$, but none in PU(n,1). It seemed tempting to look for an ergodicity proof of super-rigidity in the case of PU(n,1), $n > 1$. After such an effort failed, it became tempting to prove that super-rigidity is false in PU(n,1). That indeed turned out to be the case. In [3], I constructed lattices in U(2,1) generated by complex reflections, some of which were non-arithmetic, and for which necessarily super-rigidity fails. A few years later, I learned of an 1885 paper of E. Picard [7] constructing lattices in PU(2,1) via an entirely different method, namely, monodromy of hypergeometric functions; under certain integrality conditions, the monodromy groups are discrete and are lattices in PU(2,1). This construction generalized Schwarz's study of triangle groups in PU(1,1) ([8]). Picard's lattices seemed remarkably similar to the lattices constructed by reflections in [3], but in fact turned out to be different. Close inspection did reveal however that Picard's proof was inadequate, cf. [4]. Subsequently Deligne and I gave a correct proof of Picard's claims and generalized them to PU(n,1), that is we gave integrality conditions, called INT, which suffice to assure that the monodromy groups are discrete and lattices in PU(n,1) ([1]).

In this paper I announce a necessary and sufficient condition for the monodromy groups to be lattices in PU(n,1). This condition, called ΣINT is weaker than the previous integrality condition INT and is satisfied by the lattices constructed in [3]. Thus, both the lattices of [3] and Picard's 1885 lattices [7] are part of one larger family of monodromy groups satisfying ΣINT.

In §2, the lattices of [3] are described. In §3, the results of [1] are summarized. In §4, the results in [5] about condition ΣINT are stated. Finally, in §5, the main theorem is stated: For $d > 1$, condition ΣINT is necessary and sufficient that the monodromy group of a hypergeometric function of d variables be a lattice in PU(d,1).

The proof is sketched in §5. Full details, particularly facts used about a fundamental domain of the monodromy groups, will be presented elsewhere.

2. GROUPS GENERATED BY $\mathbb{C}$-REFLECTIONS

2.1 Let V be a vector space over $\mathbb{C}$. A *quasi-reflection* of V is a linear mapping $T:V \longrightarrow V$ with rank (T-Identity) = 1. If the matrix of T with respect to some base is $\mathrm{diag}(\lambda,1,1,..,1)$, we call T a $\mathbb{C}$-reflection with multiplier λ. The action of T on P(V), the projective space of 1-dimensional subspaces of V, is called a $\mathbb{C}$-reflection of P(V).

2.2 Coxeter has introduced diagrams for finite groups generated by $\mathbb{C}$-reflections, generalizing the diagrams used for reflections of order 2, but they can be used as well for infinite groups:

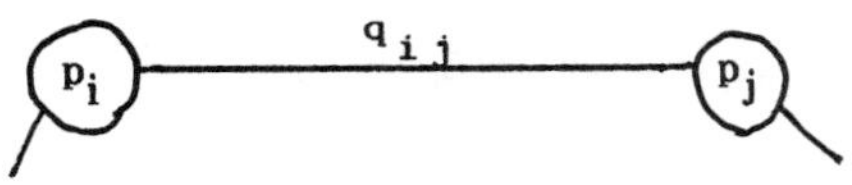

Each node is labeled with an integer p_i, and nodes i,j are joined by segments with a number q_{ij}. To such a diagram, we associated the vector space

$$V = \bigoplus_i \mathbb{C}e_i$$

on which one imposes a hermitian form satisfying

$$\langle e_i, e_j \rangle = 1 \qquad \text{for all } i$$

$$|\langle e_i, e_j \rangle| = \begin{cases} \left[\dfrac{\cos\left(\frac{\pi}{p_i} - \frac{\pi}{p_j}\right) + \cos\frac{2\pi}{q_{ij}}}{2 \sin\frac{\pi}{p_i} \sin\frac{\pi}{p_j}} \right]^{\frac{1}{2}} \\ 0, \text{ if no line joins nodes } i, j \end{cases}$$

There may be many such hermitian forms if $\arg\langle e_i, e_j \rangle$ is not specified, but all are equivalent if the graph contains no closed loops.
For each node i, define the linear map

$$R_i \colon v \longrightarrow v + \left\{ \exp\left[\frac{2\pi\sqrt{-1}}{p_i}\right] - 1 \right\} \langle v, e_i \rangle e_i;$$

R_i is a $\mathbb{C}$-reflection which preserves the hermitian form. The subgroup Γ generated by the $\{R_i\}$ will thus be a subgroup of $U(n,m)$ where $n+m = \dim V$.

2.3 In [3], I studied the lattice with Coxeter diagram

(*)

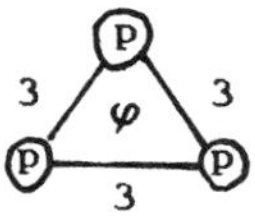

where $\langle e_1, e_2 \rangle = \langle e_2, e_3 \rangle = \langle e_3, e_1 \rangle$ and $-\varphi = \arg\langle e_1, e_2 \rangle$. For $p \leq 5$, the group generated by any two of the generating $\mathbb{C}$-reflections is finite. Set

$$t = \frac{1}{\pi} \arg \varphi^3$$

and let $\Gamma(p,t)$ denote the group with the above Coxeter diagram.

Theorem. (a) *The hermitian form of the Coxeter diagram* (*) *has signature* (2+, 1−) *for* $|t| < 3\left(\frac{1}{2} - \frac{1}{p}\right)$ *if* $p \leq 5$ (*and for* $|t| < \frac{1}{2} + \frac{3}{p}$ *if* $p \geq 6$).

(b) *For a finite number of such* t, $\Gamma(p,t)$ *is discrete.*

(c) $\Gamma(p,t)$ *is a lattice in* U(2,1) *if* $\Gamma(p,t)$ *is discrete.*

(d) *For some* t, $\Gamma(p,t)$ *is a non-arithmetic lattice in* U(2,1).

3. LATTICES ARISING FROM MONODROMY OF HYPERGEOMETRIC FUNCTIONS IN d VARIABLES

(cf. [1])

3.1 THE MONODROMY GROUP

Let $S = \{0,1,\dots,d+2\}$

P = a complex projective line

M = set of all injective maps from S to P

$P_M = \{(x,m) \in P \times M;\ x \notin m(S)\}$; for $m \in M$, set

$$P_m = P - m(S)$$

$\alpha = (\alpha_s)_{s\in S}$ a family of complex numbers indexed by S satisfying

$$\begin{cases} \alpha_s \neq 1 & \forall s \in S \\ \prod_{s\in S} \alpha_s \alpha_s = 1 & \end{cases}$$

L = a complex one-dimensional local coefficient system on P_M with monodromy α along each P_m;

that is, let L_m = restriction of L to L_m, $0 \in P_m$. Then $L_m = \mathbb{C} \times_{\pi_1(P_m,0)} \hat{P}_m$, where $\hat{P}_m$ is the simply connected cover of P_m.

We regard a constant section on $\hat{P}_m$ as a multivalued horizontal section e on P_m; i.e. if z is a local coordinate of P centered at m(s), $e(\exp(2\pi\sqrt{-1}\ \theta) \cdot z)_{\theta=1} = \alpha_s\ e(\exp(2\pi\sqrt{-1}\ \theta) \cdot z)_{\theta=0}$. That is, an element in $\pi_1(P_m, Q)$ represented by a positive loop around m(s) effects multiplication by α_s on L_m. This determines the isomorphism class of L_m uniquely, but L_m is not determined up to a unique isomorphism.

The projection $P_M \longrightarrow M$ is locally a direct product topologically. Hence $\{H^1(P_m,L_m);\ m \in M\}$ forms a flat vector bundle over M. $B(\alpha)_M = \{PH^1(P_m,L_m);\ m \in M\}$ forms a flat bundle of projective spaces over M. Hence

$$B(\alpha)_M = PH^1(P_0,L_0) \times_{\pi_1(M,0)} \hat{P}_0,$$

0 a base point in M. Let $\theta : \pi_1(M,0) \longrightarrow PGL(H^1(P_0,L_0))$ denote the action of $\pi_1(M,0)$; it results from horizontal transport of P_0.

The group Aut P operates diagonally on P^S and hence on M. Set Q = Aut P\M. Then θ descends to a homomorphism

$$\theta : \pi_1(Q,0) \longrightarrow PGL(H^1(P_0,L_0))$$

and the flat fiber bundle $B(\alpha)_M$ of projective spaces descends to a flat bundle $B(\alpha)_Q$ over Q. Similarly, L and P_m descend to bundles over Q.

<u>Def.</u> Γ_α = Image θ = the α-Hypergeometric Monodromy Group.

3.2 RELATION TO HYPERGEOMETRIC FUNCTIONS

For each $s \in S$, choose μ_s so that

$$\exp 2\pi\sqrt{-1}\ \mu_s = \alpha_s$$
$$\Sigma\mu_s = 2$$

Take $P = \mathbb{P}^1 := \mathbb{C} \cup \infty$.

Set $M_0 = \{m \in M;\ m(0) = 0,\ m(1) = 1,\ m(d+2) = \infty\}$. Then $\mathrm{Aut}\, P \times M_0 \xrightarrow{\sim} M$, and $M_0 \xrightarrow{\sim} Q$ via the quotient map. Let z denote the projection $\mathbb{P}^1 \times M_0$ to $\mathbb{P}^1$ composed with the identity coordinate of $\mathbb{P}^1$. z is then a coordinate on P_{M_0}, the pullback of P_M via $M_0 \to M$. For any $m \in M_0$, let e denote a multivalued section of L_m. Set

$$\omega_\mu(m) = \prod_{s \neq d+2} (z - m(s))^{-\mu_s}\ e\ dz$$

This defines a 1-form on P_m with coefficients in L_m, therefore a cohomology class in $H^1(P_m, L_m)$.

Lemma 1. $H^1_c(P_m, L_m) \cong H^1(P_m, L_m)$ *for all* $m \in M$.

Lemma 2. $\dim H^1(P_m, L_m) = d+1$.

Lemma 3. *The cohomology class of* $\omega_\mu(m)$ *in* $H^1(P_m, L_m)$ *is non-zero.*

Lemma 4. *The map* $m \to \omega_\mu(m)$, $m \in M_0$, *defines a holomorphic section* ω_μ *of* Q *in the flat projective space fiber bundle* $B(\alpha)_Q$.

Choose a base point $0 \in Q$. Since $B(\alpha)_Q = PH^1(P_0, L_0) \times_{\pi_1(Q,0)} \hat{Q}$, the section ω_μ of Lemma 4 defines a holomorphic multivalued map, also denoted ω_μ, of Q to $PH^1(P_0, L_0)$, i.e. a $\pi_1(Q,0)$ equivariant map $\hat{\omega}_\mu$ from $\hat{Q}$ to $PH^1(P_0, L_0)$ corresponding to a base point $\hat{0}$ above 0. We call the multivalued ω_μ the *Schwarz-Picard* map.

Now choose 0' to be the base point of M_0 given by

$$0'(s) = s \text{ for } s \neq d+2$$
$$0'(d+2) = \infty$$

We fix as base point 0 of Q_0 the point corresponding to O'.

Let $\hat{L}$ denote the inverse (or dual) of L. Then there is a perfect pairing of $H^1(P_0,L_0) \times H_1^{lf}(P_0,\check{L}_0) \to \mathbb{C}$ (lf = locally finite). As a base of $H_1^{lf}(P_0,L_0)$ one may take $\{C_1,\ldots,C_{d+1}\}$ where C_s is the interval from 0'(s) to 0'(s+1), $s \in \{1,\ldots,d+1\}$. Then $\{\langle\hat{\omega}_\mu(\hat{q}),\ C_s\rangle;\ s = 1,\ldots,d+1\}$ are homogeneous coordinates of $\hat{\omega}_\mu(\hat{q})$ where

$$\langle\hat{\omega}_\mu(\hat{q}),C_s\rangle = \int_{\hat{q}C_s} \prod_{s\neq d+2} (z-m(s))^{-\mu_s}\, dz;$$

here $\hat{q} \in \hat{Q}$ is a point in $\hat{Q}$ above $q \in Q$, m is the point in M_0 corresponding to q, and $\hat{q}C_s$ is the horizontal transport of C_s corresponding to the path from $\hat{0}$ to $\hat{q}$. Each of the integrals (*) is, by some definitions, a hypergeometric function of the d variables m(2),...,m(d+1).

Example: d = 1.

$$\int_{\hat{q}(C_1+C_2)} z^{-\mu_0} \quad (z-1)^{-\mu_1}(z-m(2))^{-\mu_2}dz$$

$$= \int_1^\infty z^{a-c}(z-1)^{c-b-1}(z-x)^{-a}dz$$

$$= \int_0^1 u^{b-1}(1-u)^{c-b-1}(1-ux)^{-a}du \ \ (\text{set } u = z^{-1})$$

$$= \frac{\Gamma(b)\Gamma(c-b)}{\Gamma(c)}\ F(a,b,c;x).$$

3.3 THE INVARIANT HERMITIAN FORM

Assume $|\alpha_s| = 1$ for all $s \in S$. Then $\hat{L} = \bar{L}$, and cup product $H^1(P_0,L_0) \times H_c^1(P_0,\bar{L}_0) \to \mathbb{C} = H_c^2(P_0,\mathbb{C})$ yields a skew-hermitian form ψ on H^1, which is preserved by the action of

$\pi_1(Q,0)$. Then $\frac{1}{2\pi\sqrt{-1}}\,\psi$ is an invariant hermitian form.

Lemma 5. *Choose* μ'_s *so that* $0 < \mu'_s < 1$, $e^{2\pi\sqrt{-1}\mu'_s} = \alpha_s$ *for each* $s \in S$. *Then the signature of the hermitian form is* $(\Sigma\mu'_s - 1, \Sigma(1-\mu'_s)-1)$.

A L_m-valued differential 1-form on P_m expressed as $\prod_{m(s)\neq\infty} (z-m(s))^{-\mu_s}\, f\, e\, dz$, where f is a meromorphic function on P, and e is a multivalued section of L_m, is called *of the first kind* if and only if $\int_{P-m(s)} \omega \wedge \bar{\omega} < \infty$.

<u>Def.</u> $H^{1,0}(P_m,L_m)$ = the vector space of forms of the first kind.

$H^{0,1}(P_m,L_m)$ = complex-conjugate of $H^{1,0}(P_m,\bar{L}_m)$.

Lemma 5 above says:

If $\sum_{s\in S} \mu'_s = 2$, then $\dim H^{1,0} = 1$.

(i.e. $\langle\omega,\omega\rangle \leqslant 0$ if $\omega \in H^{1,0}$).

The B_Q bundle section ω_μ lies in $PH^{1,0}(P_q,L_q)$ for all $q \in Q$. Thus the hypergeometric functions and the Schwarz-Picard map describe the variation of holomorphic structure on P_q via the variation of $PH^{1,0}(P_q,L_q)$ in $PH^1(P_q,L_q)$ as q varies in Q!

Furthermore, $\hat{\omega}_\mu : Q \to B^-(\alpha)_0$ = Image in $PH^1(P_0,L_0)$ of $\{h; \langle h,h\rangle < 0\}$ = the negative ball, and the α-hypergeometric monodromy group lies in $PU(H^1(P_0,L_0), \frac{1}{2\pi\sqrt{-1}}\,\psi)$.

3.4 Γ_α IS GENERATED BY $\mathbb{C}$-REFLECTIONS

As a base for $H_1^{lf}(P_m,\hat{L}_m)$, $m \in M$, one can take any union of two disjoint trees T_1, T_2 embedded in P_m so that the vertices of

T_i are in a subset $m(S_i)$ of $m(S)$ with $S = S_1 \cup S_2$. Then the 1-cells of T_i with coefficients in $\hat{L}_m$ form a base $\{C_1,\ldots,C_{d+1}\}$ for $H_1^{lf}(P_m,L_m)$.

Given any $s,t \in S$, let $\gamma_{s,t}$ in $\pi_1(M,0)$ denote a path in which t comes near s, makes a positive turn around s and returns to its original position. Take as base in $H_1^{lf}(P_0,\hat{L}_0)$ a base coming from the union of two disjoint trees $T_1 \cup T_2$ with T_1 consisting of a single 1-cell having vertices at $0(s)$ and $0(t)$. Clearly $\theta(\gamma_{s,t})$ acts trivially on all the homogeneous coordinates except those coming from the 1-cell of T_1. Thus $\theta(\gamma_{s,t})$ is a quasi-reflection on $PH^1(P_0,L_0)$: if $\alpha_s \cdot \alpha_t \neq 1$, it is a $\mathbb{C}$-reflection with multiplier $\alpha_s\alpha_t$.

Note: Set $M_\infty = \{m \in M;\ m(d+2) = \infty\}$. Then $\pi_1(M_\infty,0)$ is the colored algebraic braid group of d+2 strings in $\mathbb{R}^3$ and maps onto the monodromy group Γ_α.

We restrict ourselves now to families $\mu = (\mu_s)$, $s \in S$, satisfying $0 < \mu_s < 1$ for all $s \in S$, $\sum_{s\in S} \mu_s = 2$. Such μ are called *admissible*. We say μ satisfies condition INT if

$$\mu_s + \mu_t < 1 \text{ implies } (1 - \mu_s - \mu_t)^{-1} \in \mathbb{Z}.$$

We say μ satisfies condition ΣINT if there exists a subset $S_1 \subset S$ such that

$$\mu_s = \mu_t \text{ for all } s,t \in S_1$$

and

$$\mu_s + \mu_t < 1 \text{ implies } (1 - \mu_s - \mu_t)^{-1} \in \begin{cases} \frac{1}{2}\mathbb{Z} & \text{if } s,t \in S_1 \\ \mathbb{Z} & \text{otherwise} \end{cases}$$

We set $\Gamma_\mu = \Gamma_\alpha$ where $\alpha_s = \exp 2\pi\sqrt{-1}\,\mu_s$ for $s \in S$.

Theorem 1. (Deligne-Mostow). *If* μ *satisfies* INT, *then* Γ_μ *is a lattice in* PU(d,1).

The case d = 1 is due to Schwarz.

The case d = 2 was stated by Picard but his proof is incomplete.

Note: If μ satisfies INT, Γ_μ is not commensurable with

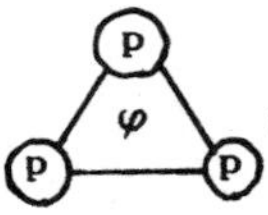

if p is odd.

4. CONDITION ΣINT (cf. [5])

Theorem 2. *If* μ *satisfies* ΣINT, *then* Γ_μ *is a lattice in* PU(d,1).

Note: If μ satisfies ΣINT, it satisfies ΣINT for a unique maximum subset S_1. In this case let Σ = Perm S_1. Then the action of Σ on P^S yields actions on M and on $B(\alpha)_M$ which descend to actions of Σ on Q and on $B(\alpha)_Q$.

Let Q' denote the subset of Q in which Σ acts freely; necessarily, $\operatorname{codim}_{\mathbb{C}}$ Q - Q' $\geqslant$ 1. The bundle map $B(\alpha)_Q \to Q$ descends to a bundle map $B(\alpha)_{Q/\Sigma} \to Q/\Sigma$, and the flat structure of $B(\alpha)_Q$ over Q yields a flat structure of $B(\alpha)_{Q'/\Sigma}$ over Q'/Σ. Thus we get a monodromy homomorphism

$$\theta_\Sigma : \Pi_1(Q'/\Sigma,\ 0) \to \operatorname{Aut} B(\alpha)_0.$$

Set

$$\Gamma_{\mu\Sigma} = \operatorname{Im}\, \theta_\Sigma$$

$\Gamma_{\mu\Sigma}$ may be regarded as an extension of Γ_μ and the index of Γ_μ in $\Gamma_{\mu\Sigma}$ divides the order of Σ.

Theorem 3. *Let* d = 2, *and let* S_1 = {0,1,2}. *Then there is a bijective correspondence*

$$\Gamma_{\mu\Sigma} \longleftrightarrow \Gamma(p,t) \longleftrightarrow \text{Coxeter diagram}$$

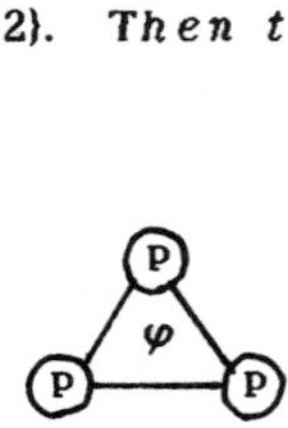

(recall $t = \frac{1}{\pi} \arg \varphi^3$) where μ ranges over 5 tuples satisfying condition ΣINT with Σ = permutation group of S_1.

The correspondence $\mu \leftarrow (p,t)$ is given by:

$$\mu = (\mu_0, \mu_1, \mu_2, \mu_3, \mu_4) \text{ with } \mu_0 = \mu_1 = \mu_2 = \frac{1}{2} - \frac{1}{p},$$
$$\mu_3 = \frac{1}{4} + \frac{3}{2p} - \frac{1}{2}t, \ \mu_4 = \frac{1}{4} + \frac{3}{2p} + \frac{1}{2}t.$$

Conversely, $p = (\frac{1}{2} - \mu_0)^{-1}$, $t = \mu_4 - \mu_3$. Moreover, $\Gamma_{\mu\Sigma} \supset \Gamma(p,t)$ and

$$[\Gamma_{\mu\Sigma} : \Gamma(p,t)]$$

$$= \begin{cases} 1 & \text{if } 3 \nmid \gcd((1 - \mu_0 - \mu_3)^{-1}, (1 - \mu_0 - \mu_4)^{-1}) \\ 3 & \text{if } 3 \mid \gcd((1 - \mu_0 - \mu_3)^{-1}, (1 - \mu_0 - \mu_4)^{-1}) \end{cases}$$

$$[\Gamma_{\mu\Sigma} : \Gamma_\mu] = \begin{cases} 6 & \text{if } p \text{ is even} \\ 1 & \text{if } p \text{ is odd.} \end{cases}$$

5. THE MAIN THEOREM

Theorem 4. *Assume* $d>1$ *and let* μ *be admissible. If* Γ_μ *is a discrete group in* PU(d,1), *then it is a lattice and* μ *satisfies condition* ΣINT.

Note: In case $d = 2$, there are 44 5-tuples satisfying conditions ΣINT for 38 of which card $S_1 \geqslant 3$. That is, most Γ_μ have extra symmetry.

<u>Sketch of the proof of Theorem 4</u>

If for any $s,t \in S$, $\mu_s + \mu_t < 1$, then the d+3 tuple μ can be contracted to a d+2 tuple ν on replacing μ_s, μ_t by $\mu_s + \mu_t$. This contraction ν gives rise to a monodromy group acting on the sub-ball of $B(\alpha)_0^-$ fixed pointwise by a $\mathbb{C}$-reflection $\theta(\gamma_{st})$, where γ_{st} results from a loop of s around t. If Γ_μ is discrete, so also is

Γ_ν. If for any 3 distinct elements a, b, c of S, $\mu_a + \mu_b + \mu_c < 1$, then the 4-tuple $\mu_a, \mu_b, \mu_c, 1-\mu_a-\mu_b-\mu_c$ is called a *Schwarz quadruple*. The Schwarz quadruples correspond to geodesic triangles on the standard 2-sphere S^2 with the property that the group generated by rotations about the vertices through double the angles generate a finite group. The angles of geodesic triangles have been listed by Schwarz and from Schwarz's list one derives the list of Schwarz quadruples; apart from the case of dihedral groups, there are 14 Schwarz quadruples.

Next, one determines all admissible 4-tuples μ with Γ_μ discrete; the μ of this kind which do not satisfy condition ΣINT fall into three families, described in the list below:

type	C_2	C_3	C_4
μ_0	$\frac{1}{2}\left(\frac{1}{2} - \frac{1}{q}\right)$	$\frac{1}{3} - \frac{1}{q}$	$\frac{1}{2} - \frac{2}{q}$
μ_1	$\frac{3}{2}\left(\frac{1}{2} - \frac{1}{q}\right)$	$2\left(\frac{1}{3} - \frac{1}{q}\right)$	$\frac{1}{2} - \frac{2}{q}$
μ_2	$\frac{1}{2}\left(\frac{1}{2} + \frac{1}{q}\right)$	$\frac{1}{3} + \frac{1}{q}$	$\frac{1}{2} + \frac{1}{q}$
μ_3	$\frac{3}{2}\left(\frac{1}{2} + \frac{1}{q}\right)$	$2\left(\frac{1}{3} + \frac{1}{q}\right)$	$\frac{1}{2} + \frac{3}{q}$
triangle group	$[2, q, \frac{q}{2}]$	$[3, q, \frac{q}{3}]$	$[q, q, \frac{q}{4}]$

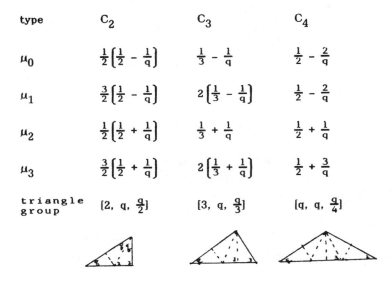

By repeated use of contraction and the limited possible Schwarz quadruples, one is reduced to the question of discreteness of the group Γ_μ with

$$\mu = \left(\frac{1}{2} - \frac{1}{p}, \frac{1}{2} - \frac{1}{p}, \frac{1}{2} - \frac{1}{p}, \frac{1}{6} + \frac{1}{p}, 2\left(\frac{1}{6} + \frac{1}{p}\right)\right).$$

$$p = 15, 24, 42$$

all of whose contracted 4-tuples correspond to discrete groups. We have by Theorem 3,

$$\Gamma_\mu = \Gamma(p,t)$$

with $t = \frac{1}{6} + \frac{1}{p}$. Define q by: $\frac{1}{p} + \frac{1}{q} = \frac{1}{6}$: then $q = 10, 8, 7$ according as $p = 15, 24, 42$ respectively.

We now require some information about a fundamental domain for $\Gamma(p,t)$ analogous to the information in [3] for the case $p \geq 5$ (cf. [6]).

Consider 5 points {0,1,2,3,4} placed on the 2 sphere S^2 with 0,1,2 equally spaced on the equator and 3,4 placed on the north and south poles respectively. Set

$$A_i = \theta(\gamma_{3i}),\ A_i' = \theta(\gamma_{4i}) \quad (i = 0,1,2)$$
$$B_0 = \theta(\gamma_{1,2}),\ B_1 = \theta(\gamma_{2,0}),\ B_2 = \theta(\gamma_{0,1}),\ B_0' = \theta(\gamma_{3,4}),$$

where the path $\gamma_{3,4}$ is selected so that B_0' commutes with B_1 and B_2, i.e. the loop $\gamma_{3,4}$ of 3 around 4 does not meet the equator between (2,0) or (0,1). Let J denote a positive rotation about the polar axis of $\frac{2\pi}{3}$ radians, i.e. $J(0) = 1$, $J(1) = 2$. Then take $B_i' = J^iB_0J^{-1}$ $(i = 1,2)$.

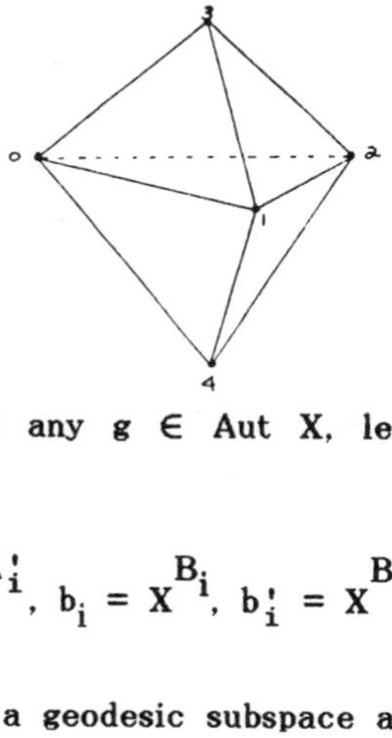

Set $X = B(\alpha)_0^-$, and for any $g \in \mathrm{Aut}\ X$, let X^g denote the fixed point set of g. Set

$$a_i = X^{A_i},\ a_i' = X^{A_i'},\ b_i = X^{B_i},\ b_i' = X^{B_i'} \quad (i = 0,1,2)$$

Each of these subsets is a geodesic subspace and is a complex 1-ball. (Each $\theta(\gamma_{st})$ is a $\mathbb{C}$-reflection on the projective 2-space $B(\alpha)_0$ and has a fixed point set consisting of a point and a line; the fact that the line rather than the point lies in the negative ball utilizes the fact that $\mu_s + \mu_t < 1$ for all $s,t \in S$.) Schematically, the twelve lines are arranged as in the accompanying diagram; all intersecting pairs of these complex lines are orthogonal.

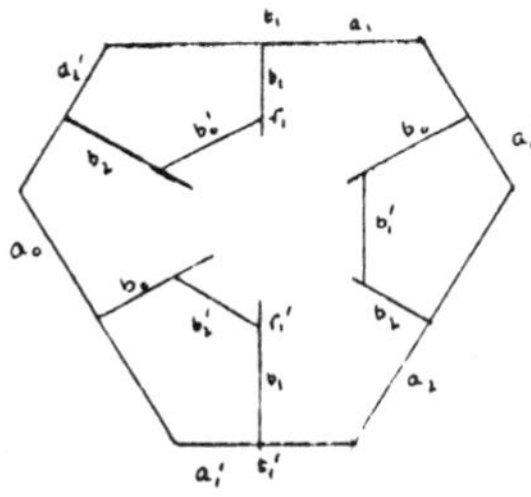

Set $t_1 = a_1 \cap b_1$, $t_1' = a_1' \cap b_1$, $r_1 = b_1 \cap b_0'$, $r_1' = b_1 \cap b_2'$. In the complex geodesic line b_1, the four points r_1, r_1', t_1, t_1' form a geodesic quadrilateral; by calculations similar to those in [3], one shows that its angles are as indicated in the diagram (cf. [6]):

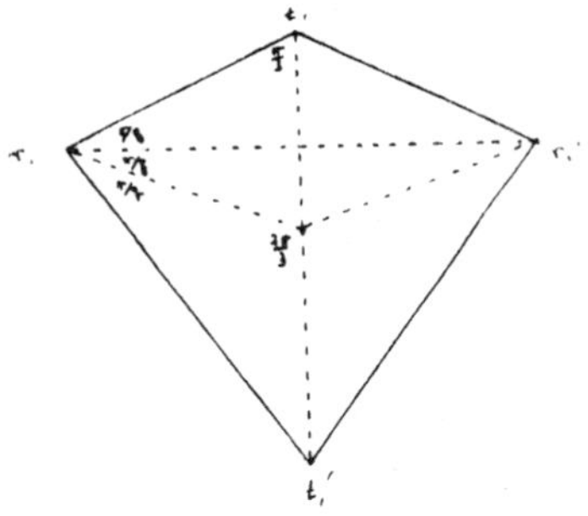

$B_0' B_1 = B_1 B_0'$ implies that $B_0' b_1 = b_1$. Inasmuch as $1 - \mu_3 - \mu_4 = 1 - \left(\frac{1}{6} + \frac{1}{p}\right) - 2\left(\frac{1}{6} + \frac{1}{p}\right) = \frac{1}{2} - \frac{3}{p} = 3\left(\frac{1}{6} - \frac{1}{p}\right) = \frac{3}{q}$, and B_0' is a $\mathbb{C}$-reflection with multiplier $(\alpha_s \alpha_t)^{-1}$ by 2.3, we see that B_0' affects a rotation in b_1 around r_1 through an angle $\frac{6\pi}{q}$. Choosing k so that $3k \equiv 1 \bmod q$, we see that $B_0'^{\,k}$ is a rotation through an angle $\frac{2\pi}{q}$. Hence $B_0'^{\,k}(r_1') = t_1'$. Since b_2' is the orthogonal complement to b_1 at r_1', we find $B_0'\ b_2' = a_1'$, the orthogonal complement to b_1 at t_1'.

To complete the proof of Theorem 4, we need the following two results, the first of which is proved in [9].

Lemma 6. *Let* [m,n,p] *denote the triangle subgroup of* PGL(2,ℝ) *generated by three generators* R_1, R_2, R_3 *with relations* $R_1^m = R_2^n = R_3^p = R_1R_2R_3 = 1$. *Then* [2,p,q] *is a maximal lattice in* PGL(2,ℝ) *except for* $p = q$ *or* $p = 2q$.

Lemma 7. *The stabilizer of* b_2' *in* $\Gamma_{\mu\Sigma}$ *is the triangle group* [2,p,q] *where* $\frac{1}{p} + \frac{1}{q} = \frac{1}{6}$; *the stabilizer of* a_1' *in* $\Gamma_{\mu\Sigma}$ *is the triangle group* [2,3,p].

The proof, given in [6], is based on computations similar to those in [3].

Inasmuch as [2,3,p] is a maximal lattice in PGL(2,ℝ) for $p > 7$ and inasmuch as PGL(2,ℝ) $\cong$ Aut a_1' and $\Gamma_{\mu\Sigma} \cap$ Aut a_1' contains triangle groups [2,3,p] and [2,p,q] with $(p,q) = (15,10), (24,8)$, or $(42,7)$ by Lemma 6, it follows by Lemma 5 that $\Gamma_{\mu\Sigma}$ is not discrete. Proof of Theorem 4 is now complete.

BIBLIOGRAPHY

[1] Deligne, P. and Mostow, G.D., Monodromy of Hypergeometric Functions, and Non-lattice Integral Monodromy, Publications, IHES, (to appear, 1985).

[2] Greenberg, L., Maximal Fuchsian Groups, Bull. Amer. Math. Soc. v. 69 (1963) pp. 569-573.

[3] Mostow, G.D., On a Remarkable Class of Polyhedra in Complex Hyperbolic Space, Pac. J. Math, v. 86 (1980), pp. 171-276.

[4] Mostow, G.D., Existence of Non-arithmetic Monodromy Groups, Proc. Nat. Acad. Sci. v. 78 (1981) pp. 5948-5950.

[5] Mostow, G.D., Generalized Picard Lattices Arising from Half-integral Conditions, Publ. IHES (to appear, 1985).

[6] Mostow, G.D., A Fundamental Domain for Monodromy Groups of Hypergeometric Functions (to appear).

[7] Picard, E., Sur les Functions Hyperfuchsiennes Provenant des Series Hypergeometrique de Deux Variables, Ann. ENS III 2 (1885) pp. 357-384.

[8] Schwarz, H.A., Ueber Diejenige Falle in Welchen die Gaussische Hypergeometrische Reihe eine Algebraische Function Ihres Viertes Elementes Darstellt, J. fur Math. t 75 (1873) pp. 292-335.

[9] Singerman, D., Finitely Maximal Fuchsian Groups, J. London Math. Soc. 2, (1972) pp 29-38.

INDUCED BUNDLES AND NONLINEAR WAVE EQUATIONS

Irving E. Segal

Dedicated to George W. Mackey

We tell the story of the happy life together of induced bundles, invariant wave equations, and scattering transformations, in the universal cosmos. Also how the first two of these are coming to live together on a more equal footing.

The scattering transformation S of a general class of nonlinear wave equations - i.e., the ultimate effect of the nonlinearity, modulo that of the linear terms, from 'infinitely early' to 'infinitely late' times, as conventionally modelled, - here comes out like apple pie. More specifically, S is essentially the action on the section space of the induced bundle in question of the (forward) generator of the infinite cyclic center of the causal automorphism group of the universal cosmos **M**. Thus, general solutions of suitable wave equations on Minkowski space-time $\mathbf{M_0}$, such as e.g., the scalar equation $\Box\varphi + \varphi^3 = 0$, extend canonically and smoothly beyond times $\pm\infty$ to all of **M**, in which $\mathbf{M_0}$ canonically imbedded as an open causal manifold. Scattering in $\mathbf{M_0}$ is thereby simplified and extended, e.g., to equations such as those of Yang-Mills. Coordination between scattering and bound state aspects is improved, and the Banach space on which S acts smoothly is typically larger and simpler, among other incidental benefits.

Philosophically, this work is an application of Mackey's Imprimitivity Theorem and theory of induced representations, as adapted in collaboration with S.M. Paneitz, primarily in the context of the homogenous space $\mathbf{M} = \widetilde{SU}(2,2)/\widetilde{P}$, where P denotes the Poincaré group, together with scale transformations (thus 11-dimensional), and the tildes denote universal covers.

1. Introduction. In the theoretical physical literature there has been an almost mystical dedication to the finding of the 'right' wave

equation in various connections. The equation was looked on as a kind of Holy Grail or magic potion, which once in hand, would be the key to the elucidation of a whole network of mysteries. In particular, in fundamental physics one typically derived the transformation properties of the fields in question from invariance properties of the equations. This is quite understandable in view of the history of the Maxwell equations, the elucidation of whose transformation properties was a major influence on the development of special relativity, and the Dirac equation, which more than any other single factor led to the strong bond between group representation theory and theoretical particle physics.

Indeed, the approach of Wigner[1)], Mackey[2)], and others, to relativistic wave equations correlated linear such equations very closely with unitary representations of the Poincaré group. Beyond this, the development of internal symmetry groups enhanced the near identity of physically relevant wave equations and unitary representations of suitable types of overall symmetry groups. The result was a major clarification and simplification of the treatment of elementary particles of great importance, but in its naive form the approach had essential limitations. First, there was a great plethora of groups and representations thereof, and it became even harder to find the 'right' group and representation thereof than a 'right' equation. Second, the formulation of local interactions between particles required an explicit presentation of wave functions on space-time, rather than as vectors in a relatively abstract representation space. The linear group representation associated with an elementary particle model is, to be sure, a very important feature of the model, but it is nevertheless descriptive of only part of its theoretical physical personality. The spatio-temporal labelling of the vectors in the representation space plays an essential part in the formation of local interactions, - locality being a crucial principle without whose imposition there would be a hopeless abundance of possibilities, - and is of course not given by the equivalence class of the associated group representation.

Such considerations led, somewhat over a decade ago, to our program[3,4,5)] attempting to model elementary particles as section spaces of homogeneous vector bundles, the homogeneous space in question being the theoretical physical space-time. Primitive and general physical notions such as causality, stability, non-anthropocentrism, and the 4-dimensionality of space-time as observed suggested mathematical constraints and thus led to a quasi-axiomatic and theoretically conservative approach to the mathematical modelling of elementary particles. This subsumed existing theoretical practice and represented a modern mathematical synthesis and extension of it. In effect, it replaced a somewhat ovesimplified conception of an elementary particle model as an irreducible group representation by the conception of it as an induced representation, inclusive of the corresponding imprimitivity system (physically interpretable as representing regions in space-time), consistently with our initiative some decades earlier, but with the greater specifity resulting from the causality and other constraints indicated.

More specifically, an elementary particle species is represented[6,7,8] as a subspace of the section space of a bundle transforming according to the causal group (i.e., group of all causality-preserving transformations) on a 4-dimensional causal manifold (i.e., manifold endowed with a convex cone field, physically interpretable as the future directions at each point). Postulating homogeneity and isotropy for the causal group, and global causality in the form of the absence of closed time-like curves (i.e., such as whose tangents point into the future at each point), in accordance with theoretical tradition and physical indications, there is a unique maximal 'cosmos', as we call such a space-time model, denoted here as **M**. Succinctly, **M** is the homogeneous space $\tilde{\mathbf{G}}/\tilde{\mathbf{P}}$, where **G** denotes SU(2,2), **P** denotes the Poincaré group extended by the inclusion of the scale transformations, or space-time magnifications, and the '~' denotes the universal cover. All other possibilities are contained in **M** as open submanifolds, whose causal groups are the subgroups of $\tilde{\mathbf{G}}$ leaving invariant the submanifold in question; and wave equations on the submanifolds in question typically extend in a natural way to **M**. The canonical local inclusions of Minkowski and the de Sitter space times in **M** are visible from the formulation of **M** as the universal cover of the local group of causal transformations (after globalization) near a point in a causally locally Minkowskian space-time, modulo the isotropy subgroup; and of the other possibilities as similar homogeneous spaces of various local subgroups of the full local group of causal transformations, satisfying the indicated homogeneity and isotropy constraints. In particular, Minkowski space $\mathbf{M}_0$ is canonically contained in $\mathbf{M}$, relative to any given origin (physically interpreted as the point of observation), and is the only possibility apart from $\mathbf{M}$ itself that is globally the direct product of 'time' and 'space' factors, where 'time' and 'space' are definable naturally in terms of the given causal structure.

These ideas have been developed in collaboration with a number of people, and especially Stephen Paneitz, whose work has been crucial. Their viability has been increasingly confirmed as their development has at the same time led in turn to some modifications in outlook. This work is much too extensive to be summarized here, and I think it will be more intelligible and interesting to focus on how it can be used to deal with a particular down-to-earth issue, - invariant wave equations on Minkowski space, both linear and nonlinear, - in a cogent way. Particularly surprising are the results on nonlinear questions. Thus, general solutions of the prototypical causally invariant wave equation $\Box\varphi + \varphi^3 = 0$ not only exist globally in $\mathbf{M}_0$ but can be continued smoothly beyond times $\pm\infty$. The result is to reduce the scattering theory for such equations to a finite temporal propagation issue in a compact space, for which there is a physically interesting and mathematically effective group-theoretic solution. In particular, the idea of scattering will be seen to extend naturally to equations such as those of Yang-Mills, to which conventional theory does not apply due to the absence of an invariant 'free' part, and to nonlinear generalized equations paralleling the derivation of linear equations from the composition series of the section spaces of induced bundles, as detailed in earlier work by Paneitz and myself. To the limited extent that nonlinear quantized field theory exists in $\mathbf{M}_0$, the same ideas apply there, and they even extend in part to linear wave equations with external sources, as in Lax-Phillips type obstacle scattering.[9]

2. $\mathbf{M}_0$ as an invariant chunk of $\mathbf{M}$. The canonical imbedding

of $\mathbf{M}_0$ into $\mathbf{M}$ is generally useful and important for what follows, so I will be a bit explicit, although nothing but analytic geometry is involved.

$\mathbf{M}_0$ is to denote Minkowski space-time as a smooth causal manifold, i.e., one endowed with a smooth closed convex cone field. In this sense $\mathbf{M}_0$ does not differ from the space $\mathbf{H}(2)$ of all 2×2 complex hermitian matrices, identified with its tangent space, and the given cone at a general point A consisting of all A+H, where H is an arbitrary positive semidefinite matrix. There is a well-known theorem of Alexandrov and Ovchinnikova, rediscovered and given a simpler proof by Zeeman, that any 1-1 transformation of $\mathbf{M}_0$ onto itself that preserves the relation $x \ll y$, defined as meaning that y-x is positive semidefinite (taking for specificity the $\mathbf{H}(2)$ formulation), is necessarily affine. This implies that the causal group of $\mathbf{M}_0$ is the scale-extended 11-dimensional Poincaré group (inclusive also of space reversal, or 'orthochrononous'). Thus $\mathbf{M}_0 = \tilde{\mathbf{P}}/\tilde{\mathbf{L}}$, where $\tilde{\mathbf{L}}$ denotes the universal cover of the scale-extended homogeneous Lorentz group; or, $\tilde{\mathbf{P}}_0/\tilde{\mathbf{L}}_0$, where the '0' subscript denotes the subgroups of codimension 1 obtained by deleting the scale transformations.

Relativistic physics, especially as treated in Wigner's classic work on representations of the Poincaré group,[1)] and developed from a general point of view by Mackey,[10)] can be described as treating elementary particle species as irreducibly invariant unitarizable subspaces (possibly infinitesimal rather than global, i.e., occurring in direct integral, rather than direct sum decompositions) of bundles induced from representations of $\tilde{\mathbf{L}}$. Here the bundle point of view is not really needed, and may appear as an unduly sophisticated affectation; it is merely implicit in the work of Wigner and Mackey, who proceed in a more direct manner.

But for dealing with wave equations on alternative space-times, or conformally invariant wave equations, the bundle viewpoint is natural and useful, if not quite essential. The extendability in a canonical way of basic wave equations to the universal cosmos $\mathbf{M}$ and developments from this have the consequence that it is quite illuminating and in substantial part sufficient to treat wave equations in $\mathbf{M}$, which moreover, avoids secondary issues that must be suppressed for brevity in any case. Thus, Maxwell's equations,[7] the Yang-Mills equations,[11] etc. extend to **M**, and **M** is the maximal space-time to which they canonically do so. In addition to the earlier descriptions of **M**, it can be defined as the universal cover of the conformal compactification $\bar{\mathbf{M}}$ of $\mathbf{M}_0$ studied by Veblen, Penrose, Todorov, and many others with theoretical physical interest. This, however, displays $\mathbf{M}_0$ as canonically imbedded in an (infinitesimally, but not globally) causal manifold that is <u>covered by</u>, rather than <u>contained in</u> **M**. For our purposes it will be important to display $\mathbf{M}_0$ as covariantly (with respect to $\tilde{\mathbf{P}}$) imbedded in **M**, and in addition to be more explicit than earlier abstract indications. In particular, the explicit causal structure in **M** is needed, although it could be defined abstractly as the unique nontrivial one that is invariant under $\tilde{\mathbf{G}}$ (apart from the distinction between past and future, i.e., causal orientation).

Because **G** is locally isomorphic to both O(2,4) and SU(2,2), one has one's choice of the stereographic projection, adapted to a Lorentzian 4-dimensional 'sphere', or the Cayley transform (one procedure generalizes to O(2,n) and the other to SU(p,p), which are never isomorphic when n exceeds 4); it is more convenient to use the Cayley tranform, since the action of $\tilde{\mathbf{G}}$ on **M** then is more familiar. As an infinitesimally causal manifold, i.e., without the stipulation that there be no closed time-like loops, but only a given convex cone field, $\bar{\mathbf{M}}$ is identical to U(2) with the translation-invariant cone field that assigns to the identity I the cone of positive semidefinite matrices in <u>**H**</u>(2), regarded in the usual way as the Lie algebra of U(2), made self-adjoint by multiplcation by i. The Cayley transform carrying a generic element H of **H**(2) into $[I+iH/2][I-iH/2]^{-1}$ is a causal isomorphism of $\mathbf{M}_0$ into U(2). The connected causal group of U(2) is SU(2,2) modulo its center, acting by transforming a general element U into $(AU+B)(CU+D)^{-1}$ if the element g of SU(2,2) takes the form $g = \begin{bmatrix} A & B \\ C & D \end{bmatrix}$, where A,B,C,D are 2 × 2 matrices. The 15-dimensional Lie algebra of all causal (or conformal, the two notions are equivalent) smooth vector fields on $\mathbf{M}_0$, which is isomorphic to su(2,2) or o(2,4), is thereby given a global action on U(2), which is covariant (or equivariant, as may be more common in the mathematical literature) with the imbedding of $\mathbf{M_0}$ into U(2) indicated. Only an 11-dimensional subalgebra of this Lie algebra integrates to global one-parameter groups of transformations on $\mathbf{M_0}$, namely that of **P**. The four remaining dimensions correspond to local causal transformations and are spanned by vector fields whose one-parameter groups become singular on $\mathbf{M_0}$, although not in $\bar{\mathbf{M}}$.

The universal cover of U(2) is $R^1 \times SU(2)$, - note that SU(2) is as an invariant Riemannian manifold identical to S^3, and it follows by general considerations that $\tilde{\mathbf{G}}$ acts globally on $R^1 \times S^3$ in extension of the indicated action of **G** on U(2), i.e., in a manner covariant with the covering transformation. The maximal linear form (or 'linearizer') of **G** is **G** itself, and an explicit form for the action of $\tilde{\mathbf{G}}$ on **M** is not known and probably too complicated to be generally useful; it will suffice for present purposes to work on U(2) in conjunction with covering considerations. Note however that the imbedding of $\mathbf{M_0}$ into U(2) is quite different from the canonical imbedding of $\mathbf{M_0}$ into **M** that is obtained locally by factoring U(2) in a neighborhood of I as locally $R^1 \times S^3$, and extending this canonically and smoothly by simple connectivity to all of $\mathbf{M_0}$. It is the latter imbedding that is needed here, and the closure $\bar{\mathbf{M}}_0$ of $\mathbf{M_0}$ in **M** is quite distinct from $\bar{\mathbf{M}}$; thus the compactification involves roughly the adjunction of a single light-cone at infinity, which infinity is neither past nor future but simultaneously both essentially, while the closure involves the adjunction of two light-cones, one in the finite past and one in the infinite future.

For an explicit description of how $\mathbf{M_0}$ fits into **M**, we must give up the covariance of the imbedding **i**, according to which **i**(gx) = g'**i**(x) for arbitrary x in $\mathbf{M_0}$ and g in $\tilde{\mathbf{P}}$, where g' denotes the action of g on all of **M**. Here $\tilde{\mathbf{P}}$ represents the isotropy subgroup for $\pi \times I$, in $\tilde{\mathbf{G}}$. $\mathbf{M_0}$ is just the orbit under $\tilde{\mathbf{P}}$ of the point of observation, $0 \times I$, in **M**, and its topological boundary consists of the future light cone in **M** with vertex $-\pi \times I$ together with the past light cone with vertex $\pi \times I$; these two vertices cover the same point of $\bar{\mathbf{M}}$, namely the matrix $-I$ in U(2). $R^1 \times S^3$ with its natural Lorentzian structure $dt^2 - ds^2$, where t is the R^1 component, in radians, while ds is the element of arc length on S^3, is known as the 'Einstein Universe' [12], having been proposed by Einstein early on as a model for the large-scale gravitational structure of the universe. For any maximal essentially compact subgroup of $\tilde{G}$, there is such an Einstein metric invariant under this subgroup $\tilde{K}$, so there is nothing unique about it, but it is here convenient to choose a particular one, and our description will be essentially $\tilde{K}$-covariant; the conjugacy of all $\tilde{K}$'s then displays the essential invariance of the present considerations. Using the usual Minkowski coordinates x_j (j = 0,1,2,3), x_0 is the Minkowski time and $(x_1^2+x_2^2+x_3^2)^{1/2}$ the Minkowski distance; in distinction to these, t will be called the 'Einstein time' and the distance from the point I of observation on S^3 the 'Einstein distance', and denoted s. In these terms, $\mathbf{M}_0$ is imbedded in $\mathbf{M}$ as the set of all points for which $|t| + s$ is less than π, and its boundary consists of the double cone $|t| + s = \pi$. The open cone defined by $0 < t \leqslant \pi$, $s = \pi - t$, and $0 > t \geqslant -\pi$, $s = \pi + t$ make up the entire boundary apart from the region at spatial infinity, $t = 0$, $s = \pi$, which represents the antipode to the point of observation, at the time of observation.

A particularly interesting and relevant transformation in $\mathbf{M}$ is that which transforms $t \times \mathbf{V}$, where $\mathbf{V}$ is now in SU(2), into $(t + \pi) \times -\mathbf{V}$; although this transformation $\mathfrak{z}'$ appears to depend on the $\tilde{\mathbf{K}}$ in question, it is in fact, the same for all $\tilde{\mathbf{K}}$, being a central element of $\tilde{G}/Z_2$, where Z_2 is a central subgroup of $\tilde{G}$ whose action on $\mathbf{M}$ is trivial, and not presently relevant (although its action on spin bundles is consequential). $\mathfrak{z}'$ is the image modulo Z_2 of the generator $\mathfrak{z}$ of the infinite cyclic center of $\tilde{G}$, or more precisely that one of the two generators which advances, rather than diminishes, the Einstein time. The union of the $\mathfrak{z}^{-n}\mathbf{M}_0$ (n = 0, ±1,...), which are disjoint, is dense in $\mathbf{M}$. In the natural action of $\tilde{G}$ on $\bar{\mathbf{M}}$, and thereby on $\mathbf{M}_0$, the action of $\mathfrak{z}$ is trivial, indeed $\bar{\mathbf{M}}$ could be described as the quotient of $\tilde{G}$ modulo the subgroup $\tilde{\mathbf{P}}\mathbf{Z}$, where $\mathbf{Z}$ denotes the center, isomorphic to $Z_2 \times Z_\infty$, of $\tilde{G}$. But the action of $\mathfrak{z}$ on $\bar{\mathbf{M}}_0$, i.e., the closure in $\mathbf{M}$ of $\mathbf{M}_0$, is quite nontrivial: the cone C_- in the past, defined by the relation $s - t = \pi$, $-\pi \leqslant t < 0$, is carried into the cone C_+ in the future, defined by the equation $s + t = \pi$, $0 < t \leqslant \pi$.

The key purely geometrical point in the relevance for

scattering in $\mathbf{M}_0$ of analysis in $\mathbf{M}$ may be stated as

THEOREM 1. *The space-like surfaces* $x_0 = c$ *in* $\mathbf{M}_0$ *converge in the Hausdorff topology on closed sets, after imbedding in* $\mathbf{M}$ *and closure by adjunction of the point* $0 \times -I$ *at spatial infinity to* $C_\pm$, *as* c *tends to* $\pm\infty$, *with the same point adjoined.*

In particular the data on $C_\pm$ given by restriction of a smooth function on $\mathbf{M}$, or on a neighborhood of $\bar{\mathbf{M}}_0$, is the limit of the data given by restriction of the function to the respective space-like surfaces in $\mathbf{M}_0$, again with the point at spatial infinity added. Note that this does not mean that under temporal translation $x_0 \longrightarrow x_0+r$, data on the Minkowski space-like surfaces $x_0=c$ converge to data on $C_\pm$ as $r \longrightarrow \pm\infty$. Indeed, every point in $\mathbf{M}_0$ converges to $(\pi, 0)$, so that the data converge to the values at these two points as $r \longrightarrow \pm\infty$. It is essential to use the Einstein parametrization involving the coordinates t and s; for $t > 0$, e.g., the data converge as $t \longrightarrow \pi - s$ to data on C_+, and similarly for C_-.

The application of Theorem 1 to wave equations depends on the extendability of wave equations from $\mathbf{M}_0$ to a neighborhood in $\mathbf{M}$, and this, in turn, becomes fully intelligible only from the induced bundle standpoint, to which I now turn.

3. Bundle-invariant wave equations. Wave equations are commonly presented in the context of structure above and beyond that of a bundle, - a specific class of coordinates, or local parallelization, element of differential geometric structure, etc. However, a pure induced bundle standpoint greatly clarifies the inter-relations of wave equations in alternative space-times, and the applicability of the canonical imbedding of $\mathbf{M}_0$ into $\mathbf{M}$, which may be called the 'conformal connection'. The case of the wave equation itself is probably the best place to start, this special case being treated as an illustration of general considerations.

If R is a given finite-dimensional representation of the isotropy group of a point in $\mathbf{M}$, $\tilde{G}$ acts canonically on the smooth section space, by a representation that will be denoted $R^{\tilde{G}/\tilde{P}}$, $\tilde{P}$ being the isotropy subgroup. The simplest case is that in which the representation space **R** of R is 1-dimensional, in which case R(g) necessarily has the form $R(g) = \lambda^w$, where w is a constant called the 'conformal weight', and g is the product of the scale transformation $x_j \longrightarrow \lambda x_j$, and a transformation in the unimodular Poincaré group $\tilde{\mathbf{P}}_0$. A scalar field of weight w on **M** is just a section of the corresponding induced bundle. When w = 1, and only then, the section space **S** of the bundle admits an interesting invariant subspace that may be correlated with solutions of the wave equation in the following way. On the one hand, it consists precisely of all sections annihilated by the differential operator $dU(\Box_f)$, where U = $R^{\tilde{G}/\tilde{P}}$ and $\Box_f$ ('f' for 'flat') denotes the element of the enveloping algebra of the Lie algebra of $\tilde{G}$, $T_0^2-T_1^2-T_2^2-T_3^2$, where T_j is the element of the Lie algebra that acts on $\mathbf{M_0}$ as $\partial/\partial x_j$. Note that this operator acts on all of **M**, and not merely on the submanifold $\mathbf{M_0}$, on which it coincides with the usual D'Alembertian. On the other hand, it can equally be described as the space of sections annihilated by $dU(\Box_c)+1$ ('c' for 'curved'), where $\Box_c$ denotes the element of the enveloping algebra of the Lie algebra of $\tilde{G}$, $X_0^2 -X_1^2 -X_2^2 -X_3^2$, X_j being the generator of **K** that acts on $\mathbf{M_0}$ as a vector field that coincides with T_j at the origin. Bundle-invariance is exemplified by the fact that $dU(\Box_f) = F(dU(\Box_c+1)$ where F is a fixed smooth funciton, given explicitly below.

It should be noted that $\Box_c+1$ does not quite correspond to the usual wave operator in the Einstein Universe, which fails to satisfy Huygens' principle, to define (by its null space) an invariant subspace of **S**, and otherwise to be appropriate. There is considerably more interesting structure, but to avoid secondary details, I will summarize. The solution manifold of the wave equation, whether in its conventional form in $\mathbf{M_0}$, or its slightly modified form in **M**, can be directly characterized as an irreducibly invariant subspace of **S** under $\tilde{G}$. This does not recover the wave operator itself, - indeed, there are a variety of them, - but does give the temporal evolution that is the main purpose of the wave operator to define; this evolution is simply a special case of the $\tilde{\mathbf{G}}$-transformation properties. (That this evolution is causal in the sense associated with hyperbolic partial differential equations, and relative to the causal structure given in **M** does however, involve the wave operator and its hyperbolicity, and is not clearly visible from the group theory; nor is the 'smallness' of the representation, in the sense that sections in the subspace are determined by their restrictions to a space-like surface, together with the restrictions of one derivative.) Moreover the scalar case is quite typical; the situation is similar for the Dirac, Maxwell, and so-called higher spin equations, which (in their 'massless' forms) correspond to irreducibly invariant, unitarizable subspaces of the section spaces of bundles induced from other representations of $\tilde{\mathbf{P}}$ that are trivial on the translations, and are holomorphic on the homogeneous Lorentz group cover, realized as the complex group SL(2,C), and have a uniquely determined conformal weight determined by this representation of SL(2,C).

In addition, every 'normalizable' solution of the wave equation on $\mathbf{M_0}$, - i.e. having finite norm relative to the unique Poincaré-invariant inner product in the solution manifold, - extends uniquely to a distribution on **M** satisfying the wave equation there, and the norm is invariant under all of $\tilde{\mathbf{G}}$, and not merely $\tilde{\mathbf{P}}$, - but for this to be true it is essential to take a bundle invariant formulation. As a group, i.e., the universal cover of U(2), any bundle over **M** is parallelizable by left or right translation, and computations are greatly facilitated by presentations of sections in parallelized form. Similarly for bundles over $\mathbf{M_0}$. These two parallelizations are, however, quite distinct, as a consequence of which if a section is represented in the 'curved' (**M**-parallelization) by φ, as a function to the representation space **R** of the inducing representation R, its representative ϕ in the 'flat' ($\mathbf{M_0}$-parallelization) is different, having the form $\phi = L\varphi$, where L is a linear operator on **R** that depends on the point of $\mathbf{M_0}$ in question. This is true of all induced bundles; in the case of the scalar bundle of conformal weight, 1, L consists of multiplication by the function p on **M**, p having the form $\frac{1}{2}(\cos t + \cos s)$, where t and s are respectively the Einstein time and distance. Thus, p vanishes on $C_{\pm}$, but not on $\mathbf{M_0}$. There is a similar function for every conformal weight, but only for $w = 1$ does the function F, which takes the form p^2, exist.

Similar phenomena take place in bundles induced from arbitrary holomorphic representations of SL(2,C), extended to $\tilde{\mathbf{P}}$, e.g., in the spinor case there is an association with the Dirac equation, in the vector case with the Maxwell equations, etc.; the corresponding solution manifolds appear as irreducibly invariant subspaces of the respective section spaces, and lacking invariant complements in the section space as a whole; the function F becomes a matrix. It is probable that there is a general group-theoretic phenomenon here, but as yet only an aspect of the scalar case has been treated in a more general way by Kostant[13]. On the other hand, the methods used by Jakobsen and Vergne[14] and by Speh[15] in their studies of representation theory for $\tilde{\mathbf{G}}$, although not bundle-theoretic in the present sense, appear in part generalizable as well as quite informative concerning the case treated. It should also be noted that the first correlation of invariant massless wave equations with unitary representations of $\tilde{\mathbf{G}}$, by Leonard Gross[16], also avoided the use of curved spaces such as **M** and worked entirely in $\mathbf{M_0}$.

In summary, in terms of the functions obtained by the respective flat and curved parallelizations, the situation for the wave equation (similar to that for arbitrary 'spin' equations) is that any distribution ϕ on $\mathbf{M_0}$ that satisfies the usual wave equation, and has finite Lorentz-invariant norm, has the form $\phi = p\varphi \mid \mathbf{M_0}$, for a unique function φ on **M** satisfying the modified wave equation on **M** as $R^1 \times S^3$, $(\Box_c+1)\varphi = 0$. Such φ form an irreducibly invariant subspace of the section space, on which $\tilde{\mathbf{G}}$ acts unitarily, so that for an arbitrary smooth function ϕ on $\mathbf{M_0}$, $\Box_f\ (p\phi) = p^3(\Box_c+1)\phi$, in consequence of the relation between the two parallelizations and the indicated bundle-invariance of the wave operator. This is purely a relation between differential operators, and could be verified directly by brute force, or derived by Riemanninan geometry considerations as was done originally by Ørsted[17]. There are similar relations for other induced bundles, as earlier.

It is surprising that a solution of the wave equation in $\mathbf{M_0}$, whose generic behavior at infinity displays little explicit regularity, should be smoothly and uniquely extendable to a solution of a natural extension of the equation on a space that is compact in the spatial direction, such as **M**, and effectively compact in the time direction in view of the periodicity of the extensions with period 2π (invariance under ζ^{-2}). In effect, the solutions are being extended beyond times $\pm\infty$, but the Minkowski time coordinate becomes singular there (the vector field corresponding to T_0 vanishes), and must be replaced by the Einstein time coordinate. It is more surprising that the same is true for nonlinear invariant wave equations, such as the Yang-Mills equations, as shown in work with Y. Choquet-Bruhat and S.M. Paneitz[11]. In order to develop the implications for scattering theory here, this essentially general result will be stated for the simplest non-trivial case. The equation $\Box_f \phi + \phi^3 = 0$ in $\mathbf{M}_0$ has been studied as regards global existence and the Cauchy problem, by Jörgens[18] and others, and as regards scattering theory by Strauss[19] and others; moreoever, the equation is conformally invariant, under the infinitesimal action of the conformal group on $\mathbf{M}_0$, of the appropriate weight (1).

THEOREM 2. *Given sufficiently regular Cauchy data for the equation* $\Box_f \phi + \phi^3 = 0$ *in* $\mathbf{M}_0$ *(e.g., in Sobolev spaces of sufficiently high order, and remaining in these spaces after multiplication by polynomials of sufficiently high order), there exists a unique solution* φ *of the equation* $(\Box_c + 1)\varphi + \varphi^3 = 0$ *on* **M** *such that* $\phi = p\varphi \mid \mathbf{M}_0$.

Conversely, given any solution φ *of this equation in* **M** *of finite Einstein energy,* $\phi = p\varphi \mid \mathbf{M}_0$ *is a finite Minkowski energy solution of the equation given in* $\mathbf{M}_0$. *Moreover, such solutions* φ *exist for arbitrary finite Einstein energy data (solution in a strong, but not necessarily strict abstract evolutionary sense, - 'mild' solution in the sense of Kato), and form a* $\tilde{\mathbf{G}}$*-invariant Banach manifold. If the Cauchy data are infinitely differentiable, then the corresponding* ϕ *is in the*

Schwartz space and the foregoing applies.

The proof is, in its basic aspects, a considerable simplification of the similar results for the Yang-Mills case.[11] What is distinctive about the scalar case is the existence of a development of scattering theory, which involves, as usual, a separation of the total hamiltonian into a 'free' and 'interacting' part. No such invariant separation exists in the case of the Yang-Mills and other complex equations, but the induced bundle approach will show the existence of a natural scattering transformation. This will be validated by consideration of the scalar nonlinear case, the essential lines of the argument being clearly independent of the spin.

4. Invariant scattering theory. To do so, recall the general concept of scattering. Suppose X and Y are vector fields on a manifold **L** that generate smooth one-parameter groups of diffeomorphisms of **L**; X may be thought of as the generator of the 'free' motion, and Y as that for the 'interacting' or perturbed motion, but the formal treatment is symmetric in X and Y. Let **X** and **Y** denote the classes of all mappings from R^1 to **L** of the forms $e^{tX}p$ and $e^{tY}p$ respectively, p being arbitrary in **L**; p is the 'Cauchy datum' at time 0. For each real t, there is a natural map from **X** to **Y**, defined as carrying any given x in **X** into the y in **Y** having the same Cauchy datum at time t; the limit of this map, say P_t, if it exists, as $t \longrightarrow \pm\infty$, is the wave operator $W_\pm$ from **X** to **Y**. If W_-^{-1} exists, the scattering transformation S, from **Y** to **Y** is defined as $W_+W_-^{-1}$. Similarly for the scattering transformation from **X** to **X**. There is no special mathematical reason for the indicated limits to exist, and in the case of the finite-dimensional linear manifold it is easily seen that generically they do not, even for linear transformations.

From the standpoint of applications, it is the free-to-free S-transformation that is most convenient, corresponding to the universal reductionist tendencey to try to analyze states of complex systems in terms of the simpler states of hypothetical 'free' systems. But formally there is complete symmetry between the two S-transformations, and for mathematical purposes, it will be seen that the interacting-to-interacting S-transformation is more fundamental. Indeed, this is the case in fundamental physical principle also, but there is no known way to correlate the states of theoretical nonlinear systems with empirically observed ones except by the use of the free states as labels.

Thus, in the conventional quantum mechanical case of a free hamiltonian H_0 and a total hamiltonian H, one may define $W_\pm$ as the limits as $t \longrightarrow \pm\infty$ of $e^{-itH_0} e^{itH}$. It is obvious that one has trouble with the limit if H has nontrivial point spectrum. Reversing the role of H_0 and H does not materially affect the problem when H_0 and H are self-adjoint operators in a Hilbert space. It will be seen that in the nonlinear case being treated here, the use of the interacting-to-interacting S-transformation eliminates the difficulty connected with the distinction between 'bound' states, i.e., eigenvectors for H or their nonlinear analogue, and 'scattering' states, which have to be treated separately in the usual analysis.

THEOREM 3. *For the equation* $\Box_f\phi + \phi^3 = 0$ *in* $\mathbf{M}_0$*, the wave operators* $W_\pm$ *relative to the free equation* $\Box_f\phi = 0$*, are as follows, in their interacting to free forms, and in terms of the extended equation given by Theorem 2, for sufficiently smooth data.*

$W_\pm\ \varphi$ *is the solution of the free wave equation* $(\Box_c+1)\varphi_\pm = 0$ *whose data on the light cone* $C_\pm$ *are those for* φ.

The proof uses Theorem 1 to show that the data on the space-like surfaces $x_0 = c$ in $\mathbf{M}_0$ for the given wave function ϕ, when transferred to the 'curved' wave function φ in **M**, converge as the time becomes infinite. However, instead of holding the Minkowski space position fixed and letting c become infinite (which would lead to the values at $\pm\pi \times I$, which vanish because p vanishes at these points), it is necessary to use the natural analogue in **M**, i.e., to let the Einstein time t approach $\pi - s$ (for $c \longrightarrow -\infty$). Indeed, φ is smooth and has well-defined traces on $C_\pm$. To propagate these data on C_+, say back to time 0 by the free equation is to solve the free equation with these data on C_+, a Goursat rather than a Cauchy problem known to have a unique solution. Note that although $C_\pm$ are outside $\mathbf{M}_0$ they are causally equivalent to conventional light-cones in $\mathbf{M}_0$, and the wave equation itself is unaffected by such a transfer back into $\mathbf{M}_0$.

In terms of the action of $\tilde{G}$ on the section space of the weight 1 scalar bundle, the scattering transformation that derives from Theorem 3 has a very simple representation. Let $\Gamma(g)$ denote the action of the element g of $\tilde{G}$ on the section space.

COROLLARY 3.1. *The scattering transformation for the equation* $\Box_f\phi + \phi^3 = 0$ *in* $\mathbf{M}_0$, *for sufficiently regular Cauchy data at one time (e.g., as in Theorem 2), exists and apart from a phase factor (-1 to be specific) is* $\Gamma(\mathfrak{I})$ *in its interacting-to-interacting form (expressed in the first instance as an action on the bundle over* $\mathbf{M}$, *and then restricted to* $\mathbf{M}_0$). *It is thus a restriction of the* $\tilde{G}$-*invariant transformation* $\Gamma(\mathfrak{I})$ *applicable to all solutions of the equation on* $\mathbf{M}$, *of finite Einstein energy.*

The main question here is of the existence and character of $W_+^{-1}W_-$. Now W_- takes a solution φ of the nonlinear equation on $\mathbf{M}$, determines its data on the cone C_-, and solves the Goursat problem for the free equation with these data, obtaining the solution φ of the free equation. What W_+^{-1} does is to carry this free solution into a solution of the nonlinear equation having the same data on the cone C_+. But $\mathfrak{I}$ carries C_- into C_+ and carries a free solution into its negative. Thus, the transform of φ by $\Gamma(\mathfrak{I})$, multiplied by -1, will be such a solution of the nonlinear equation, which is uniquely determined by its data on the cone.

In the case of the Yang-Mills equation, which appears prototypical for more complex equations, there is no invariant separation of the total hamiltonian into a free and interaction part, - indeed the hamiltonian takes exactly the same form as in the free

Maxwell equations, apart from the matrix values of the fields and the need to take traces thereof. Nevertheless the action Γ of the group $\tilde{G}$ is well-defined, and the relations between solutions on $\mathbf{M}_0$ and $\mathbf{M}$ are similar to those for the scalar case, apart from the complications (which are substantial but presently irrelevant) required to deal with the two gauges involved. This suggests the

DEFINITION. *The scattering transformation for the Yang-Mills equation, or any other conformally invariant equation that extends together with generic solutions from* $\mathbf{M}_0$ *to,* $\mathbf{M}$ *is* $\Gamma(\zeta)$.

The scattering transformation S thus is conformally invariant, symplectic, regular, etc., and embodies the intuitive idea of scattering as carrying the field in the infinite past into the infinite future. (The infinite past and future are here interpreted in Minkowski space terms, rather than Einstein terms, but that should be quite sufficient for direct microscopic applications, which empirically involve time intervals that are generally short, or at least finite in Minkowski terms. This comment should be ignored in a purely mathematical reading, it is made only to avoid undue oversimplification of the physical interpretations.)

Not to oversimplify the mathematics either, note that scattering on $\mathbf{M}_0$ cannot be conformally invariant in its most complete sense simply because $\mathbf{M}_0$ is not conformally invariant. The Cauchy data treated are required to vanish sufficiently strongly at spatial infinity, which is tantamount to requiring their extendability from R^3 to S^3 smoothly so as to vanish at the antipode of the point of observation, - a condition that cannot be conformally invariant, and can reasonably be made so only by working in $\mathbf{M}$, it would appear. Thus, to show the conformal invariance of scattering on $\mathbf{M}_0$ for the indicated scalar equation, when defined (depending on the wave functions and conformal transformations in question) in $\mathbf{M}_0$ would appear to be quite difficult by an analysis entirely within $\mathbf{M}_0$, although it is an immediate deduction from Corollary 3.1.

5. Scattering for other types of wave equations. Other types of scattering besides the classical (i.e., unquantized), conformally invariant cases just considered are of considerable 'practical' importance. In the quantized theory, the temporal evolution is linear, nonlinearity appearing only in the analytic form of the interaction hamiltonian. In obstacle and potential scattering the evolution is again linear. In addition, if the power ϕ^3 is replaced by a higher odd power (oddness being required for a positive definite energy, required in turn for global solubility of the equation with some generality), covariance is only with respect to the Poincaré group, which does not include $\mathfrak{r}$. Nevertheless, it remains the case that the scattering transformation, when it exists, is still given by an action of $\mathfrak{r}$.

Consider first the case of a quantized field, taking as earlier the wave equation as representative prototype. Denoting such fields by boldface letters and otherwise as in the classical case, the quantized free field $\boldsymbol{\phi}$ in $\mathbf{M}_0$ may be regarded as a quasi-operator-valued function on $\mathbf{M}_0$ ('quasi') in that the values $\boldsymbol{\phi}(x)$ at a point x of $\mathbf{M}_0$ are continuous sesquilinear forms on the domain of all infinitely differentiable vectors with respect to the Einstein energy operator in its natural metric topology; interestingly, one must turn to the energy in $\mathbf{M}$, rather than that in $\mathbf{M}_0$, which has a smaller energy and correspondingly larger domain of infinitely differentiable vectors, on which $\boldsymbol{\phi}(x)$ is not defined; in informal physical language, there are 'infra-red divergences', which are eliminated by using the Einstein in place of the Minkowski energy operator as the dominant one). This field satisfies the same equation as earlier, $\Box_f\boldsymbol{\phi} = 0$, and other relations characteristic of quantized fields: canonical commutation relations, positivity of the energy operator on the free field state vector space $\mathbf{B}$, etc. In the particle, or Fock-Cook[20] representation, $\mathbf{B}$ appears as the direct sum of all symmetrized powers of the 'single-particle' Hilbert space $\mathbf{H}$, which may be described as the Poincaré irreducibly invariant space of all normalizable classical solutions ϕ of the wave equation $\Box_f\phi = 0$ in $\mathbf{M}_0$ (say, positive-energy and complex valued).

As earlier, the Hilbert space $\mathbf{H}'$ consisting of the conformally irreducibly invariant space of all normalizable solutions of the wave equation $(\square_c + 1)\varphi = 0$ on **M** is unitarily equivalent to **H**, via the map $\varphi \longrightarrow p\varphi \mid M_0$. This induces a corresponding unitary equivalence between **B** and the corresponding space for the field over **M**, which has the effect of carrying $\phi(x)$ into $p(x)^{-1}\varphi(x)$, formally just as for the classical fields. The quantized wave equation field over M_0 is thus identical to the restriction of that over **M**, apart from a relabelling of the quantities involved and the factor p. For other fields, – Maxwell, etc., – the situation is entirely similar.

When an interaction is introduced, the situation is formally similar, but can naturally be made mathematically sound only to the extent that the interacting field is such. At the present time, the mathematical theory for the formal quantized equation $\square\varphi + \phi^3 = 0$ is largely non-existent, so that no rigorous counterpart to Theorem 3 can be given for quantized fields. Nevertheless, one expects the existence in a suitably generalized sense of a representation Γ of $\tilde{G}$ by unitary operators on the field Hilbert space **B**, or some generalized form thereof. On this purely formal basis, the scattering transformation can be identified with $\Gamma(\mathfrak{r})$ as earlier; but one can do somewhat better.

A central result of (necessarily, at this point, heuristic) quantum field theory is a perturbative expansion for the S-operator of the putative quantized field $\square\phi + g\phi^3 = 0$, where g is a positive constant, as a power series in g, of the form $S = 1 + 1/4\, g \int_{M_0} :\phi(x)^4:\, d_4x + O(g^2)$. Here $:\phi(x)^4:$ is a generalized power known as the Wick power, and treated by Gårding and Wightman[21], myself[22] and other authors in a rigorous mathematical way. The indicated expression for S forms the basis of the Feynman diagram technique for 'practical' quantum field theory.[23] An application of the 'conformal connection' described earlier relating analysis on M_0 to analysis on **M**, together with a method due to Skovhus Poulsen[24], shows that the leading nontrivial term $\int_{M_0} :\phi(x)^4: d_4x$ is convergent (as an integral of an operator-valued distribution), to an essentially self-adjoint operator in

B, on the domain of all infinitely differentiable vectors for the Einstein energy operator in B.[25] Thus, the same ideas as earlier are helpful also in the quantized case in establishing properties of S, and are likely to provide a basis for the analysis of the precise mathematical relation between the so-called 'Heisenberg field', which is the analogue of the classical nonlinear $\tilde{G}$-invariant field earlier treated, and the 'interaction representation' formulation of the dynamics which describes it, as briefly indicated, in terms of free fields.

Returning to the classical case, equations need not be conformally invariant for the general method to apply. Although there is then no representation Γ associated with the full group $\tilde{G}$, the analogue of $\Gamma(\mathfrak{r})$ may be well defined. More specifically, if it is possible to extend solutions ϕ of the equation given in M_0 to solutions φ of a corresponding equation in M, where $\phi = p\varphi$, and if this equation is smoothly soluble in M, or indeed only in the closure of M_0 in M, then S is definable as earlier in terms of the data for the solution on the cones C_- and C_+ in the 'infinite past' and 'future'. Thus, the respective 'in' and 'out' free fields for the given solution ϕ of the nonlinear equation on M_0 are of the form $p\varphi_\pm$, where $\varphi_\pm$ are the solutions of the free equation on the closure of M_0 having as Goursat data the traces of φ on $C_\pm$. Taking, for example, an equation of the form $\Box_f\phi + g\phi^q = 0$ on M_0, in terms of φ on M this takes the form $(\Box_c+1)\varphi + gp^{q-3}\varphi^q = 0$, by the 'covariance' of the wave operator earlier indicated; and for higher spin equations the situation is similar. The coefficient p^{q-3} is perfectly regular on M if q is at least 3. Local solubility thus presents no problem in $\bar{M}_0$; global solubility in this region has to depend on special properties of the simple function p, but at least on a pertubative basis (for small g or small Cauchy data at one time), global existence follows by the general method earlier developed for scattering of nonlinear wave equations in M_0.

In the case of wave equation scattering, whether obstacle or potential, the same method applies in principle. In the similar cases of scattering by a potential of compact support, i.e., the equation $\Box_f\phi$

$+ V(\vec{x})\phi = 0$, where V is a smooth function of space alone having compact support, the limits on $C_{\pm}$ of the associated solution φ exist and serve as earlier to define the scattering operator. In working on **M** one loses the considerable practical advantage of applicability of Fourier analysis, but may gain a theoretical advantage, as in the treatment of trapped rays, a partial analogue to bound states. In the more general case in which V does not necessarily have compact support, which has recently been treated by Phillips,[26) there is a singularity due to the occurrence of uncompensated negative powers of p, which vanishes on $C_{\pm}$, but the singularity is not severe and limits on $C_{\pm}$ may be obtainable by a more careful analysis in **M**.

However, there is no solid physical reason to prefer $\mathbf{M_0}$ to **M** as the space-time in which to analyze wave equation scattering, and there is the mathematical advantage that an analysis in **M** should be more general than one in $\mathbf{M_0}$, in that it will deform into that analysis as the space curvature of S^3, the spatial component of **M**, is allowed to tend to zero. In practical scattering issues, the distances involved are quite infinitesimal compared to the distance scale of the universe, however estimated reasonably, and the indicated deformation is the same as taking the local region in which scattering is considered to be infinitesimal in relation ot S^3. Static obstacle scattering in $\mathbf{M_0}$ and **M** are not quite equivalent due to large-scale differences in the decomposition of space-time into time and space factors. It is extremely improbable that this could affect any practical scattering issues.

The same can be said of non-relativistic potential scattering, which can be considered to be derived from the wave equation scattering by the approximation $(m^2 - \Delta + V)^{1/2} \sim m - (\Delta/2m) + (V/2m)$. There is the same enhanced generality in taking space as a sphere S^3 of radius R, since it is probable that the results deform as R tends to ∞ into those for R^3. The principal term in the Schrödinger equation, $-i(\partial/\partial t) - (\Delta/2m)$, where t is the Einstein time and Δ is the Laplace-Beltrami operator on S^3, is, of course, not conformally invariant, but the Schrödinger equation tends to inherit properties of the wave equation, apart from its finite propagation velocity. The transference of considerations from $\mathbf{M_0}$ to **M** would only permit extension beyond Minkowski times $\pm\infty$ of an approximation to the solution $\mathbf{M}_0$, but the approximation involved here ($R \longrightarrow \infty$) may be of comparable order to that involved in the non-relativistic approximation ($c \longrightarrow \infty$).

Except in the last case, only apparently massless systems have been considered, i.e., the constant m that intervenes in conventional massive wave equations vanishes in the cases so far considered. It may, however, be construed as a fixed potential, which, as earlier seen, introduces a singularity, but not one of the greatest severity, which may therefore, be analytically manageable. On the other hand, from a fundamental position, a fixed constant mass is unnatural if Mach's principle is accepted, and represents only an approximation. The mass then represents the effect of the large-scale, distant univese, which is, of course, continually fluctuating, so that m^2 merely represents the mean value of a stochastic quantity of small dispersion. Such fluctuations can be modelled in a simple conformally invariant manner by coupling the field ϕ to an additional field ψ, whose square has expectation value m^2 in the relevant state, by equations of the form

$$\Box\phi + \psi^2\phi + \phi^3 = 0, \qquad \Box\psi + \phi^2\psi = 0$$

for example.

Finally, let me return to the question of the relation between induced bundles and nonlinear wave equations, as a chicken/egg issue. For linear wave equations there is a natural bundle-theoretic derivation as a defining equation for an irreducibly invariant positive-energy subspace of the section space of the bundle. In the nonlinear case this relation would appear to be irremediably lost, at first glance, but the role of the solution manifold as the locus of infinite-dimensional function theory in quantized field theory suggests an analogy. Taking, for example, the particularly interesting case of the Yang-Mills equations, is the ideal in the function algebra over the section space of the bundle in question that is generated by these equations (l.h.s set equal to 0) a minimal smooth conformally and gauge invariant ideal (closed in an appropriate topology, and perhaps positive energy in some sense). The association of generalized solutions of postive-energy equations with ideals in the associated function algebra [27] provides a possible basis for bringing this question down to earth.

REFERENCES

1. E.P. Wigner, On unitary representations of the inhomogeneous Lorentz group. Ann. Math. (2) **40**(1939), 149-204.

2. G.W. Mackey, Group representations in Hilbert space. Appendix to Mathematical problems of relativistic physics, I.E. Segal, American Mathematical Society, Providence, 1963, 113-130.

3. I.E. Segal, Covariant chronogeometry and extreme distances. Astron. & Astrophys. **18**(1972), 143-148.

4. I.E. Segal, H.P. Jakobsen, B. Ørsted, S.M. Paneitz & B. Speh, Covariant chronogemetry, II: Elementary particles. Proc. Natl. Acad. Sci. USA **78**(1981), 5261-5265.

5. I.E. Segal, Covariant chronogeometry and extreme distances, III: Macro-micro relations. Internatl. J. Theor. Phys. **21**(1982), 851-869.

6. S.M. Paneitz & I.E. Segal, Analysis in space-time bundles, I: General considerations and the scalar bundle. J. Functional Analysis **47**(1982), 78-142,

7. S.M. Paneitz & I.E. Segal, Analysis in space-time bundles, II: The spinor and form bundles. J. Functional Analysis **49**(1982), 335-414.

8. S.M. Paneitz, Analysis in space-time bundles, III: Higher spin bundles. J. Functional Analysis **54**(1983), 18-112.

9. P.D. Lax & R.S. Phillips, Scattering theory, Academic Press, New York, 1967.

10. G.W. Mackey, Unitary group representations in physics,

probability, and number theory, Benjamin/Cummings Co., Reading MA, 1978.

11. Y. Choquet-Bruhat, S.M. Paneitz, and I.E. Segal, The Yang-Mills equations on the universal cosmos, J. Functional Analysis **53**(1983), 112-150.

12. C. MØller, The general theory of relativity, 2nd ed., (1972) Clarendon Press, Oxford, Eng.

13. B. Kostant, Verma modules and the existence of quasi-invariant differential operators. Non-commutative harmonic analysis (Actes. Colloq., Marseille-Luminy, 1974), 101-128, Lecture Notes in Math. **466**, Springer, Berlin, 1975.

14. H.P. Jakobsen & M. Vergne, Wave and Dirac operators, and representations of the conformal group. J. Functional Analysis **24** (1977), 52-106.

15. B. Speh, Degenerate series representations of the universal covering group of SU(2,2), J. Functional Analysis **33** (1979), 95-118.

16. L. Gross, Norm invariance of mass-zero equations under the conformal group. J. Math. Phys. **5**(1964), 687-695.

17. B. Ørsted, Conformally invariant differential equations and projective geometry, J. Functional Analysis **44**(1981), 1-23.

18. K. Jörgens, Über die nichtlinearen Wellengleichungen der Mathematischen Physik, Math. Ann. **138**(159), 179-202.

19. W.A. Strauss, Decay and asymptotics for $\Box u = F(u)$, J. Functional Anal. **2**(1968), 409-457.

20. J.M. Cook, The mathematics of second quantization. Trans.

Amer. Math. Soc. **74**(1953).

21. A.S. Wightman & L. Gårding, Fields as operator-valued distributions in relativistic quantum theory. Ark. Fys. **28**(1964), 129-184.

22. I.E. Segal, Nonlinear functions of weak processes. I, J. Functional Analysis **5**(1969), 404-456; II, ibid. **6**(1970), 91-108.

23. J.D. Bjorken & S.D. Drell, Relativistic quantum fields, 1965. McGraw-Hill Co., New York.

24. N.S. Poulsen, On C^{∞} vectors and intertwining bilinear forms for representations of Lie groups. J. Functional Analysis **9**(1972), 87-120.

25. S.M. Paneitz & I.E. Segal, Self-adjointness of the Fourier expansion of quantized interaction field Lagrangians. Proc. Natl. Acad. Sci. USA **80**(1983), 4595-4598.

26. R.S. Phillips. Scattering theory for the wave equations with a short range potential. Indiana Univ. Math. J. **31**(1982), 609-639.

27. I.E. Segal, Banach algebras and nonlinear semigroups. Integral Equations and Operator Theory **4**(1981), 435-455.

COMPACT ABELIAN AUTOMORPHISM GROUPS OF INJECTIVE SEMI-FINITE FACTORS

Dedicated to Professor G.W. Mackey

Masamichi Takesaki

INTRODUCTION. After the breathtaking breakthrough in the analysis of the structure of injective factors and their automorphism groups by A. Connes during 1975/76, [3,4,5], fine structure analysis of group actions on injective semi-finite factors came into the theory of operator algebras. V. Jones completed a classification of actions of finite groups on an injective II_1-factor in his thesis, [13]. A. Ocneanu further supplied an important technical tool, called the stability lemma at infinity, [18], and proved the triviality of the non-commutative second cohomology of amenable group actions on an injective II_1-factor. Dualizing Ocneanu's result, V. Jones classified prime actions of compact abelian groups on an injective II_1-factor [14], where an action of an abelian group is said to be prime if it has the entire dual group as its Connes Γ-spectrum. These results can be viewed as a purely non-commutative theory. If we drop the assumption of primeness, then the usual ergodic theory comes into the picture. In these notes, we will present a real blend of commutative and non-commutative ergodic theories based on a recent joint work with V. Jones [15].

Research supported in part by NSF Grant 8120790.

CHARACTERISTIC INVARIANT AND MAIN THEOREM. Let $\mathcal{M}$ be a separable von Neumann algebra. Let Aut($\mathcal{M}$) denote the group of all automorphisms of $\mathcal{M}$ equipped with the topology given by the following system of neighborhoods of the identity ι:

$$V(\omega_1,\ldots,\omega_n)=\{\alpha\in \text{Aut}(\mathcal{M}):\|\omega_j\circ\alpha-\omega_j\|+\|\omega_j\cdot\alpha^{-1}-\omega_j\|<1, i\leqslant j\leqslant h\}$$

where $\{\omega_1,\ldots,\omega_n\}$ runs over all finite subsets of the predual $\mathcal{M}_*$. The normal subgroup of Aut($\mathcal{M}$) consisting of all inner automorphisms will be written Int($\mathcal{M}$). It turns out that Aut($\mathcal{M}$) is a Polish group, whilst Int($\mathcal{M}$) need not be closed. The quotient group Out($\mathcal{M}$) = Aut($\mathcal{M}$)/Int($\mathcal{M}$) is topologically highly non-trivial. The group of all unitaries in $\mathcal{M}$ will be written $\mathcal{U}(\mathcal{M})$, which is a Polish group with respect to the σ-strong operator topology. For each $u\in\mathcal{U}(\mathcal{M})$, Ad(u) denote the inner automorphism: $x\in\mathcal{M} \longrightarrow uxu^*\in\mathcal{M}$. The kernel of the homomorphism: $u\in\mathcal{U}(\mathcal{M}) \longrightarrow \text{Ad}(u) \in \text{Int}(\mathcal{M})$ is precisely the unitary group $\mathcal{U}(\mathcal{Z})$ of the center $\mathcal{Z}$ of $\mathcal{M}$. Hence, Int($\mathcal{M}$) is a Borel subgroup of Aut($\mathcal{M}$), being isomorphic to the Polish quotient group $\mathcal{U}(\mathcal{M})/\mathcal{U}(\mathcal{Z})$.

By an action of a locally compact group G on $\mathcal{M}$ we mean a continuous homomorphism α of G into Aut($\mathcal{M}$). We assume that α is faithful in the sense that $\alpha_s=\iota$ if and only if s=e, the unit of G. With an action α of G on $\mathcal{M}$, we construct the crossed product $\mathcal{M} \rtimes_\alpha G$ which is the von Neumann algebra on $L^2(G, \mathfrak{H}) = L^2(G) \otimes \mathfrak{H}$ generated by the operators:

$$(\pi_\alpha(x)\xi)(s) = \alpha_s^{-1}(x)\xi(s),\ x \in \mathcal{M},\ s \in G,$$
$$(u(g)\xi)(s) = \xi(g^{-1}s),\ g \in G,\ \xi \in L^2(G, \mathfrak{H}),$$

where $\mathfrak{H}$ is the Hilbert space on which $\mathcal{M}$ acts. When G is abelian, the dual action $\hat{\alpha}$ of the dual group $\hat{G}$ on $\mathcal{M} \rtimes_\alpha G$ is defined to the the action given by the unitary representation v of $\hat{G}$ on $L^2(G, \mathfrak{H})$:

$$(v(p)\xi)(s) = \overline{\langle s, p\rangle}\,\xi(s),\ p\in\hat{G}.$$

DUALITY THEOREM. *There exists an isomorphism* Φ *of the second crossed product* $(\mathcal{M} \rtimes_\alpha G) \rtimes_{\hat{\alpha}} \hat{G}$ *onto* $\mathcal{M} \bar{\otimes} \mathcal{L}[L^2(G)]$ *such that* $\Phi\tilde{\tilde{\alpha}}_s\Phi^{-1} = \alpha_s \otimes \rho_s$, $s \in G$ *where* ρ *is the action of* G *on* $L^2(G)$ *induced by the regular representation* ρ *of* G *on* $L^2(G)$:

$$(\rho(g)\xi)(s) = \xi(sg), \quad s,g \in G.$$

Actions α of G on $\mathcal{M}$ and β of G on $\mathcal{N}$ are said to be conjugate if there exists an isomorphism Φ of $\mathcal{M}$ onto $\mathcal{N}$ such that $\beta_s = \Phi\alpha_s\Phi^{-1}$, $s \in G$. We write this fact as $\alpha \cong \beta$. Now, a strongly continuous $\mathcal{U}(\mathcal{M})$-valued function $u: s \in G \longrightarrow u_s \in \mathcal{U}(\mathcal{M})$ is called an α-cocycle if

$$u_{st} = u_s\alpha_s(u_t), \quad s,t \in G.$$

In this case, we make a new action ${}_u\alpha$ by

$${}_u\alpha_s = Ad(u_s)\cdot\alpha_s, \quad s \in G,$$

and call it perturbation of α by u. Two actions α of G on $\mathcal{M}$ and β of G on $\mathcal{N}$ are said to be cocycle conjugate if there exists an α-cocycle u such that ${}_u\alpha \cong \beta$. We write this fact as $\alpha \simeq \beta$. We say that α and β are stably conjugate if $\alpha\otimes\rho \cong \beta\otimes\rho$, and write $\alpha \sim \beta$. The following implications can be checked easily:

$$\alpha \cong \beta \Rightarrow \alpha \simeq \beta \Rightarrow \alpha \sim \beta.$$

An action α is called centrally ergodic if α acts ergodically on the center $\mathcal{Z}$, i.e., the fixed point algebra $\mathcal{Z}^\alpha$ reduces to the scalars $\mathbb{C}$. By the usual direct integral decomposition, any action is a direct integral of centrally ergodic actions. Hence, we may and do assume the central ergodicity for our actions. In fact, every action dual to an action on a factor is centrally ergodic.

Now, let A be a fixed compact abelian group and G be the dual of A. Suppose that two actions α and β of A on $\mathcal{M}$ and $\mathcal{N}$ are stably conjugate. Let θ be an isomorphism of $\mathcal{M}\bar{\otimes}\mathcal{L}(L^2(A))$ onto $\mathcal{N}\bar{\otimes}\mathcal{L}(L^2(A))$ such that $\theta(\alpha_a\otimes\rho_a)\theta^{-1} = \beta_a\otimes\rho_a$, $a\in A$. The two actions α and β are said to have the same <u>inner invariant</u> if

$$\theta(1\otimes\int_A\rho(a)da) = \sigma(1 \otimes\int_A\rho(a)da) \quad \text{for some} \quad \sigma\in \mathrm{Aut}(\mathcal{N}\bar{\otimes}\mathcal{L}(L^2(A)))$$

commuting with $\alpha\otimes\rho$. Note $\int_A\rho(a)da$ is the one dimensional projection on $L^2(A)$ to the space spanned by the constant function 1.

As an easy consequence of the duality theorem, we can prove the following:

THEOREM 1 *i) Two actions α and β of A on factors $\mathcal{M}$ and $\mathcal{N}$ are stably conjugate if and only if the dual actions are cocycle conjugate.*

ii) Two stably conjugate actions are conjugate if and only if they have the same inner invariant.

Therefore, the classification problem of actions of A is quickly transferred to the cocycle conjugacy classification problem of centrally ergodic actions of G. Since a cocycle perturbation of an action α of G does not have any effect on the action to the center $\mathcal{Z}$, our problem includes the conjugacy classification problem of ergodic actions of G on a standard measure space, whose solution does not seem to be in sight yet. Thus, we are forced to take the classification of $\{\mathcal{Z}, G, \alpha\}$ as granted.

We are now in the situation that a centrally ergodic action α of a countable discrete group G on $\mathcal{M}$ is given. Let $N_\alpha=\alpha^{-1}(\mathrm{Int}(\mathcal{M}))$. The <u>characteristic</u> <u>invariant</u> of α will be the pair (λ, μ), modulo coboundaries, defined by the following conditions:

$$\begin{aligned} \alpha_h &= \mathrm{Ad}w_h, && h\in N_\alpha. \\ w_h w_k &= \mu(h,k)w_{h,k}, && h,k\in N_\alpha, \\ \alpha_g(w_h) &= \lambda(h,g)w_h, && g\in G,\ h\in N_\alpha. \end{aligned}$$

Thus, λ maps $N_\alpha \times G$ into $\mathcal{U}(\mathcal{Z}(\mathcal{M}))$ and μ: $N_\alpha \times N_\alpha$ $\mathcal{U}(\mathcal{Z}(\mathcal{M}))$. We denote the characteristic invariant of α by χ_α. It follows that λ and μ enjoys the following properties:

$$\mu(h,k) = 1 = \lambda(h,g) \text{ whenever } h=1 \text{ or } g=1;$$
$$\mu(h,k)\mu(hk,l) = \mu(h,kl)\mu(k,l), \quad h,k,l \in N_\alpha;$$
$$\lambda(hk,g)\lambda(h,g)^*\lambda(k,g)^* = \mu(h,k)\alpha_g(\mu(h,k)^*);$$
$$\lambda(h,gg') = \lambda(h,g)\alpha_g(\lambda(h,g')); \qquad g,g' \in G$$
$$\lambda(h,k) = \mu(k,h)\ \mu(h,k)^*.$$

If $f:N_\alpha \longrightarrow \mathcal{U}(\mathcal{M})$ is any function with $f(1)=1$, then we define the pair $\delta f = (\delta_1 f, \delta_2 f)$ by

$$(\delta_1 f)\ (h,g) = f(h)\ \alpha_g(f(h)^*), \quad g \in G, \ h \in N_\alpha,$$
$$(\delta_2 f)\ (h,k) = f(hk)\ f(h)^* f(k)^*, \quad h,k \in N_\alpha.$$

It follows that δf enjoys the above properties of (λ, μ). The set of all pairs (λ, μ) satisfying the above conditions form a multiplicative group, which we denote by $Z(G,N_\alpha,\mathcal{U}(\mathcal{M}))$. The subset consisting of all δf's is a subgroup of $Z(G,N_\alpha,\mathcal{U}(\mathcal{M}))$ denoted by $B(G,N_\alpha,\mathcal{U}(\mathcal{M}))$. The change of $\{w_h\}$ by $\{f(h)w_h\}$ will result in the multiplication of (λ, μ) by δf. Thus, the characteristic invariant χ_α is the image of (λ, μ) in the quotient group $\Lambda(G,N_\alpha,\mathcal{U}(\mathcal{M})) = Z(G,N_\alpha,\mathcal{U}(\mathcal{M})/B(G,N_\alpha,\mathcal{U}(\mathcal{M}))$.

It is easy to see that χ_α is invariant under cocycle perturbations. Thus, if α and β are cocycle conjugate then: i) the restrictions of α and β to the centers are cocycle conjugate; and, ii) $N_\alpha = N_\beta$ and there exists an isomerphism θ of $\mathcal{Z}(M)$ or to $\mathcal{Z}(N)$ conjugating the restrictions of α and β such that $\theta(\chi_\alpha) = \chi_\beta$.

Therefore, our main question is whether or not the above conditions (i) and (ii) are sufficient for cocycle conjugacy. In order to answer the question, it is clear that we must have a good control of the base von Neumann algebra $\mathcal{M}$. This forces us to consider only injective semifinite von Neumann algebras. In fact, J. Phillips showed that there are continuously many non cocycle conjugate actions of $\mathbb{Z}$

on the group von Neumann algebra $\mathcal{R}(\Phi)$ of the free group of infinite generators with trivial characteristic invariant, [20].

We thus consider injective semi-finite von Neumann algebras only. This assumption is supported by the grand result of Connes which states that the crossed product of an injective von Neumann algebra by an amenable locally compact group is always injective. In particular, the crossed product of an injective semi-finite factor by a compact abelian group is injective and semi-finite, but not necessarily a factor.

Let us now fix an injective semi-finite von Neumann algebra $\mathcal{M}$ with center $\mathcal{O}$. Suppose that α is a centrally ergodic action of a countable discrete abelian group G on $\mathcal{M}$. Let $\{X, \mu\}$ be a standard measure space such that $\mathcal{O}$ can be identified with $L^\infty(X, \mu)$. Then we get the direct disintegration of $\mathcal{M}$:

$$\mathcal{M} = \int_X^\oplus \mathcal{M}(x)d\mu(x)$$

over the center $\mathcal{O}$ so that almost every component $\mathcal{M}(x)$ is an injective semidefinite factor. Let τ be a faithful semidefinite normal trace on $\mathcal{M}$. For each $g \in G$, set

$$\rho_g(t) = (D\tau \circ \alpha_g : D\tau)_t, \quad t \in \mathbb{R}, \ g \in G.$$

Then ρ_g is a continuous one parameter unitary group in $\mathcal{O}$. Hence, ρ_g is of the form:

$$\rho_g(t) = k_g{}^t, \quad t \in \mathbb{R}, \ g \in G.$$

with a non-singular positive self-adjoint operator affiliated with $\mathcal{O}$. In other words, k_g is a postive measurable function in X such that

$$\tau(\alpha_g(a)) = \int_X \tau_x(a(x))k_g(x)d\mu(x), \quad a \in \mathcal{M},$$

where

$$\tau = \int_X^{\oplus} \tau_x d\mu(x)$$

is the direct disintegration of τ. It follows that ρ_g is a one parameter unitary group valued cocycle in the sense that

$$\rho_{gh}(t) = \alpha_h^{-1}(\rho_g(t))\,\rho_h(t), \quad g, h \in G, \ t \in \mathbb{R}.$$

The group ρ_g depends on the choice of τ, but its cohomology class $[\rho]$ does not. We call it the unitary module of the action α. In terms of $\{k_g\}$, the cocycle identity becomes

$$k_{gh}(x) = k_g(hx)k_h(x), \quad g,h \in G, \ x \in X.$$

The cohomology class $[k]$ of $\{k_g\}$ will be called the module of α, and denoted by $\mathrm{mod}(\alpha)$.

We now state the main theorem:

THEOREM 2. *Let* α *and* β *be centrally ergodic actions of countable abelian group* G *on semi finite injective von Neumann algebras* $\mathcal{M}$ *and* $\mathcal{N}$. *Two actions* α *and* β *are cocycle conjugate if and only if*

i) $\mathcal{M} \cong \mathcal{N}$;

ii) The restrictions of α *and* β *to the respective centers are conjugate;*

iii) $N_\alpha = N_\beta$;

iv) There exists an isomorphism θ *of* $\mathcal{Z}_{\mathcal{M}}$ *onto* $\mathcal{Z}_{\mathcal{N}}$ *conjugating* $\{\mathcal{Z}_{\mathcal{M}}, \alpha\}$ *and* $\{\mathcal{Z}_{\mathcal{N}}, \beta\}$ *such that* $\theta(x_\alpha) = x_\beta$, $\theta(\mathrm{mod}\,(\alpha)) = \mathrm{mod}\ \beta$.

Thanks to the result of Connes, [4], the condition (i) is not a big issue. First of all, the central ergodicity imples that $\mathcal{M}$ is of the form:

$$\mathcal{M} \cong \mathcal{R} \bar{\otimes} \mathcal{A},$$

where $\mathcal{R}$ is an injective semifinite factor and $\mathcal{A}$ is the center $\mathcal{Z}_{\mathcal{M}}$ of $\mathcal{M}$. Next, $\mathcal{R}$ is either of type I, or type II, or of type II_∞, i.e., $\mathcal{R}$ is isomorphic to $\mathcal{L}(\mathbb{K})$ for some Hilbert space, the unique hyperfinite II_1-factor $\mathcal{R}_0$, the infinite tensor product of 2 × 2 matrix algebras, or to the tensor product $\mathcal{R}_{0,1} = \mathcal{R}_0 \bar{\otimes} \mathcal{L}(\mathbb{K})$ with dim $\mathbb{K} = +\infty$. The condition (ii) is a central problem in ergodic theory.

The main point of the present paper is that the condition (iv) together with (i), (ii) and (iii) guarantees the cocycle conjugacy.

In the case that α appears as the action dual to an action of a compact abelian group on a semifinite injective factor, mod(α) must be trivial. In fact, mod(α)=1 if and only if α is a cocycle conjugate to an action dual to such a system.

APPROXIMATELY FINITE DIMENSIONAL VON NEUMANN ALGEGRAS. A von Neumann algebra $\mathcal{M}$ is by definition approximate finite dimensional if it is the σ-weak closure of an increasing union of finite dimensional*-subalgebras. Murray and von Neumann proved in the last paper of their pioneering work that an AFD factor of type II_1 is unique up to isomorphism. This was the first definite result in the structure analysis of factors. The second result was obtained by H.A. Dye [9], who proved that the factor of the group measure space construction by a finite measure preserving ergodic abelian transformation group is AFD, thus, isomorphic to the factor studied by Murray and von Neumann. In this work, Dye introduced an important concept in ergodic theory: the concept of orbit equivalence for ergodic transformation groups. Two ergodic transformation groups $\{X_1, G_1, \mu_1\}$ and $\{X_2, G_2, \mu_2\}$ are said to be orbit equivalent if there exists a non singular transformation W of X_1 onto X_2 such that $WG_1x=G_2W_1x$ for almost every $x \in X_1$. He proved that the orbit structure is unique for finite measure preserving ergodic transformation groups of polynomial growth which include all abelian infinite groups. Thus, if G is an infinite abelian group of non-singular measure preserving transformations of a standard finite measure space $\{X, \mu\}$, then $\{G,X,\mu\}$ is orbit equivalent to the irrational rotation action of $\mathbb{Z}$ on the torus $\mathbb{T}$ as well as to the translation action of the algebraic product of $\mathbb{Z}_2$ on the cartesian product of $\mathbb{Z}_2$.

After the discovery of a continum of factors of type III by R.T. Powers in '67 [21], and a classification theory of factors of infinite tensor product of finite type I factors by Araki and Woods, [1], the uniqueness question of AFD factors of type II_∞ was raised. It attracted many specialists. But the lack of intrinsic characterization of AFD algebras blocked the progress. Although every factor $\mathcal{R}$ of type II_∞ is isomorphic to the tensor product $\mathcal{R}_1 \bar{\otimes} \mathcal{B}$ of a factor of type II_1 and a factor $\mathcal{C}$ of type I_∞ isomorphic to the algebra of all bounded operators on a Hilbert space. There was no way to transfer approximating finite dimensional subalgebras of $\mathcal{R}$ down to $\mathcal{R}_1$.

On the other hand, J.T. Schwartz, [22], discovered a remarkable property of AFD-algebras, called Property (P): for any $x \in \mathcal{L}(\mathfrak{H})$ the weak convex closure of $\{uxu^*: u \in \mathcal{U}(\mathcal{M})\}$ meets $\mathcal{M}'$. It was implicitly conjectured that Property (P) might characterize the AFD property. Hakeda and Tomiyama introduced another property, called *Type* (E). [12], which is now called the *injectivity*, [4]. Namely, a von Neumann algebra is called injective if there exists a projection $\mathcal{E}$ of norm one from $\mathcal{L}(\mathfrak{H})$ onto $\mathcal{M}$. A celebrated theorem of Tomiyama states that a projection $\mathcal{E}$ of norm one from a C-*algebra A to a C*-subalgegra B enjoys the properties:

i) $\mathcal{E}$ is completely positive in the sense that $[\mathcal{E}(x_{ij})]$ is positive for every positive nxn matrix $[x_{i,j}]$ over the algebra;

ii) $\mathcal{E}(axb) = a\mathcal{E}(x)b$, $a, b \in B$, $x \in A$.

Another property, called *semi-discrete*, was further interoduced by Effros and Lance. A von Neumann algebra $\mathcal{M}$ is called *semi-discrete* if the map: $\sum_{i=1}^{n} x_i \otimes x_i' \in \mathcal{M} \otimes \mathcal{M}' \longrightarrow \sum_{i=1}^{n} x_i x_i' \in \mathcal{L}(\mathfrak{H})$ can be extended continuously to the C*-algebraic injective tensor product $\mathcal{M} \otimes_{min} \mathcal{M}'$.

A. Connes finally proved that these properties are all

equivalent to the AFD property, [4]. In many respects, the injectivity and the semi-discreteness are very easy to handle. For example, if the original algebra is injective, then it is very easy to prove that the crossed product by an amedable group is also injective. Another example is that every von Neumann subalgebra of an injective finite von Neumann algebra is trivially injective. Thus, every von Neumann subalgebra of an AFD factor $\mathcal{R}_0$ of type II_1 is AFD. In other words, $\mathcal{R}_0$ is "smallest" among all infinite dimensional factors in the sense that it is only infinite dimensional factor imbedable to any infinite dimensional factor.

ACTIONS ON SEMI-FINITE INJECTIVE FACTORS. Let $\mathcal{R}$ denote a fixed semifinite injective factor. By the work of Connes, we have only four cases:

$$\mathcal{R} \cong \begin{cases} \mathcal{M}(n;\mathbb{C}): \text{ nxn-matrix algebra}, \\ \mathcal{L}(\mathbb{K}): \text{ the algebra of all bounded operators on } \mathbb{K}, \\ \mathcal{R}_0: \text{ the AFD factor of Murray and von Neumann}, \\ \mathcal{R}_{0,1} = \mathcal{R}_0 \bar{\otimes} \mathcal{L}(\mathbb{K}). \end{cases}$$

In the case that $\mathcal{R}$ is of type I, every automorphism is inner, so that an action of G on $\mathcal{R}$ means simply a projective unitary representation. Since the regular representation of G absorbs every other projective representation, the cocycle conjugacy problem of actions of G on $\mathcal{R}$ means simply the classification problem of multipliers of G. If G is abelian, this is effectively translated to the study of the compact abelian group of all anti-symmetric bicharacters of G.

For $\mathcal{R}=\mathcal{R}_0$ and $G=\mathbb{Z}$, A. Connes classified all actions of G on $\mathcal{R}$ up to cocycle conjugacy in [3,5]. The techniques and the results of Connes prompted further progress in the classification theory. V. Jones then classified all actions of finite groups on $\mathcal{R}$ in his thesis [13] and A. Ocneanu did it for discrete amenable groups in his thesis [18] also. One can view these results as purely noncommutative ergodic theory.

Once we allow a non-atomic center in the carrier algebra, then the usual ergodic theory enters in the theory. Now, suppose that a discrete countably infinite group G acts on an injective semi-finite von Neumann algebra $\mathcal{M}$ via α. Suppose that α is centrally ergodic. We then have $\mathcal{M} \cong \mathcal{A} \bar{\otimes} \mathcal{R}$ with $\mathcal{R}$ an injective semi-finite factor. We identify $\mathcal{A}$ with $L^{\infty}(X, \mu)$ for some standard measure space $\{X, \mu\}$. The action α of G on $\mathcal{A}$ is then implemented by a non-singular action of G on $\mathcal{A}$ so that

$$(\alpha_g(a))\ (x) = a(g^{-1}x),\ g\in G,\ x\in X,\ a\in\mathcal{A}.$$

With the central decomposition.

$$\mathcal{M} = \int_X^{\oplus} \mathcal{M}(x)d\mu(x),$$

to each $(g,x)\ \in G \times X$, there corresponds an isomorphism $\alpha_{g,x}$ of $\mathcal{M}(x)$ to $\mathcal{M}(gx)$ so that

$$(\alpha_g(a))\ (gx) = \alpha_{g,x}(a(x)),\ a\in\mathcal{M},\ (g,x)\ \in G \times X.$$

In $\mathcal{G} = G \times X$, we define a groupoid structure as follows:

$$r(g,x) = gx,\ s(g,x) = x.$$
$$(gh,x) = (g,hx)\ (h,x),\ g,\ h\in G,\ x\in X.$$

It then turns out that $\mathcal{G}$ is a measured orbitally discrete groupoid. To each $x\in X$, there corresponds the isotropy group $H_x = \{g\in G\ :\ gx=x\}$. By ergodicity and the commutativity of G, H_x is independent of almost every $x\in X$. So let H denote the common isotropy group, which is in turn the kernel of the restriction of α to the center $\mathcal{A}$; hence it appears as the Connes essential spectrum of the original action of the compact abelian group on the factor in the case that α is a dual action. Let K=G/H be the quotient group. The restriction of α to the center $\mathcal{A}$ gives rise naturally to an action $\dot{\alpha}$ of K on $\mathcal{A}$, which is ergodic and free. Hence, $\mathcal{K}$ = KxX is a principal measured orbitally discrete groupoid. Thanks to the result of

Dye [9], Connes-Krieger, [7] and Feldman-Lind [10] or to the even more powerful result of Ornstein-Weiss [19] and Connes-Feldman-Weiss [6], the principal groupoid $\mathcal{K}$ is hyperfinite so that it is generated by a single ergodic non-singular transformation. Thus $\mathcal{K}$ has no cohomological obstruction at all from degree two on. This gives us the following advantage.

Whilst the short exact sequence:

$$1 \longrightarrow H \longrightarrow G \longrightarrow K \longrightarrow 1$$

does not split in the most cases, the short exact sequence of groupoids:

$$X \longrightarrow \mathcal{H} = H \times X \longrightarrow \mathfrak{G} = G \times X \longrightarrow \mathcal{K} = K \times X \longrightarrow X/G$$

does split. Hence $\mathfrak{G}$ is isomorphic to the cartesian product groupoid H $\times$ $\mathcal{K}$. This fact simplifies the whole theory drastically. Namely, the study of an action α of G splits into two parts: the "H-part" and the "K-part", and the combination of the two.

Thus, we look at first the "H-part". It consists of a field $\{\alpha_x\}$ of actions of H on fibres $\{\mathcal{M}(x) : x \in X\}$. Since each fibre algebra $\mathcal{M}(x)$ can be identified with a single injective factor $\mathcal{R}$ in a measurable fashion, the field $\{\alpha_x\}$ can be viewed as a measurable family of actions of H on the factor $\mathcal{R}$. The commutativity of G and the ergodicity of G on X then imply that the inner part $N_x = \alpha_x^{-1}$ (Int $\mathcal{R}$) is independent of $x \in X$, so that there exists a subgroup N of H such that $N = N_x$ for almost every $x \in X$. Thus, the action α of G gives rise to a map: $x \in X \longrightarrow \chi_x = \chi_{\alpha_x} \in \Lambda(H,N,\mathbb{T})$, which is measurable since $\Lambda(H,N,\mathbb{T})$ is a compact abelian group, the two combined facts of ergodicity and commutativity implies that χ is constant. Hence, the previous work of Jones [14], shows that the actions $\{\alpha_x\}$ of H are mutually cocycle conjugate. Thus, one can perturbe $\{\alpha_x\}$ into a constant action of H by a cocycle. Now, thanks to the splitting of the groupoid exact sequence, the cocycle can be extended to G as a one cocycle. Therefore, we can perturbe the whole action α of G by a cocycle of G in such a way that the resulted field of actions of H is a constant field. Therefore we come to the situation that an action of H on $\mathcal{R}$ and another action of K on $\mathcal{R}$, which commute with each other, are given. We denote the both actions by α.

Our strategy is then to work separately on the "H-part" and the "K-part". But the "H-part" and the "K-part" are, of course, not independent. The first problem is as to how we solve an obstruction for finding a cocycle of G whose restrictions to H and K agree with preassigned cocycles of H and K. The second problem is that after conjugating perturbed H-actions, we have to conjugate perturbed K-actions without disturbing the first conjugation of the H-part.

Since the first problem is a genuine cohomological obstruction, it is a serious problem. We shall solve this problem by constructing a model action which we have better control to the extent that we can kill the obstruction by an appropriate conjugation of itself to a cocycle perturbation. We shall discuss this in the next section.

For the second problem, we introduce the following Polish groups and then apply the Cohomology Lemma of Krieger and Bures-Connes-Sutherland.

Let β be an action of H on a separable factor $\mathcal{P}$. Let $\mathrm{Aut}^H(\mathcal{P})$ be the set of all pairs $(\theta, \{u_t\})$ where $\theta\in\mathrm{Aut}(\mathcal{P})$ and $u\in Z^1_\beta(H,\mathcal{U}(\mathcal{P}))$ such that

$$\theta\beta_t\theta^{-1} = \mathrm{Ad}(u_t)\beta_t,\ t\in H.$$

With multiplication:

$$(\theta_1, \{u_t\})\,(\theta_2, \{v_t\}) = (\theta_1\theta_2, \{\theta_1(v_t)u_t\}).$$

$\mathrm{Aut}^H(\mathcal{P})$ is a group. The usual $\mathrm{Aut}(\mathcal{P})$-topology on the θ-part and the pointwise $\mathcal{U}(\mathcal{P})$-convergence topology on the u-part make $\mathrm{Aut}^H(\mathcal{P})$ a Polish group. Inside $\mathrm{Aut}^H(\mathcal{P})$, the dual group $\hat{H}$ sits in a natural way as elements of the form: $(\mathrm{id}, \{p(t)\})$, $p\in\hat{H}$. Thus $\hat{H}$ is a closed central subgroup of $\mathrm{Aut}^H(\mathcal{P})$, and hence the quotient group $\mathrm{Aut}^H(\mathcal{P})/\hat{H}$ $= \mathrm{Aut}_H(\mathcal{P})$ is a Polish group. We denote the quotient map: $\mathrm{Aut}^H(\mathcal{P})$

$\longrightarrow \mathrm{Aut}_H(\mathcal{P})$ by π. Another normal subgroup of $\mathrm{Aut}^H(\mathcal{P})$ is $\mathrm{Int}(\mathcal{P})$, where $\mathrm{Ad}u$, $u \in \mathcal{U}(\mathcal{P})$, is identified with the element $(\mathrm{Ad}(u), \{u\beta_t(u^*)\})$ of $\mathrm{Aut}^H(\mathcal{P})$. The next result then gives the base of the application of the Cohomology Lemma. We refer the reader to the original paper for the proof.

LEMMA. *Let* $\mathcal{R}$ *be an injective semi-finite factor. If* ρ *is the closed subgroup of* $\mathrm{Aut}^H(\mathcal{R})$ *consisting of pairs* $(\theta, \{u_t\})$ *such that* $\mathrm{mod}\ \theta=1$, *then* $\pi(\mathrm{Int}(\mathcal{R}))$ *is a dense normal Borel subgroup of the closed subgroup* $\pi(\mathrm{P})$.

MODEL ACTIONS. We now discuss the most subtle point of the entire theory. To understand the nature of the problem and the strategy of attacking it, we consider a much simpler case.

Suppose $G = H \times K$ and α, β are two actions of G on a factor $\mathcal{P}$. Suppose that the restrictions of α and β to H and K are respectively cocycle conjugate in such a way that the conjugation of α/K and β/K does not interfere the conjugation of $\alpha|H$ and $\beta|H$. Thus, we may assume that there exist $u \in Z^1_\beta(H,\mathcal{U}(\mathcal{P}))$ and $v \in Z^1_\beta(K,\mathcal{U}(\mathcal{P}))$ and $\theta \in \mathrm{Aut}(\mathcal{P})$ such that

$$\theta\alpha_h\theta^{-1} = \mathrm{Ad}(u_h)\beta_h, \qquad h \in H.$$
$$\theta\alpha_k\theta^{-1} = \mathrm{Ad}(v_k)\beta_k, \qquad k \in K.$$

The trouble is that $w_g = u_h v_k$, $g = hk \in G$, need not be a cocycle on G. Since $\mathrm{Ad}(u_h)\beta_h$ and $\mathrm{Ad}(v_h)\beta_k$ commute, we get a bicharacter φ on $H \times K$ such that

$$u_h\beta_h(v_k) = \varphi(h,k)\ v_k\beta_k(u_h), \ h \in H,\ k \in K.$$

This bicharacter φ is a genuine obstruction for readjustment of w to obtain a cocycle on G. We have fortunately a method to overcome these difficuties. Suppose that we have an explicitly constructed action m of G such that for any bicharacter φ on $H \times K$ there exist $\theta \in \mathrm{Aut}(\mathcal{P})$, $\bar{u} \in Z^1_m(H,\mathcal{U}(\mathcal{P}))$ and $\bar{v} \in Z^1_m(K,\mathcal{U}(\mathcal{P}))$such that

$$\theta m_h \theta^{-1} = \mathrm{Ad}\bar{u}_h m_h, \quad h \in H,$$
$$\theta m_k \theta^{-1} = \mathrm{Ad}\bar{v}_k m_k, \quad k \in K.$$
$$\bar{u}_h m_h(\bar{v}_k) = \overline{\varphi(h,k)}\, \bar{v}_k m_k(\bar{u}_h), \; h \in H,\; k \in K.$$

We then compare α and m. After a preliminary conjugation of α, there exist $u \in Z_m^1(H, \mathcal{U}(\mathcal{P}))$, $v \in Z_m^1(K, \mathcal{U}(\mathcal{P}))$ and φ such that

$$\alpha_g = \mathrm{Ad}(u_h m_h(v_k)) m_g, \quad g = hk \in H \times K \text{ and,}$$
$$u_h m_h(v_k) = \varphi(h,k)\; v_k m_k(u_h).$$

We now further conjugate α by the above θ so that

$$\theta \alpha_g \theta^{-1} = \mathrm{Ad}(\theta(u_h m_k(v_k))\; \bar{u}_h m_h(\bar{v}_k)) m_g,$$

and $hk \longrightarrow \theta(u_k m_h(v_k)) \bar{u}_h m_h(\bar{v}_k)$ is indeed a cocycle, hence we are through.

Our task is therefore to construct an action m of G with a preassigned characteristic invariant χ and module and the above extra properties.

The choice of χ and the module δ cause a natural restriction on the type of $\mathcal{R}$. For example, if δ is non-trivial, then $\mathcal{R}$ must be of type II_∞; if $N \mathcal{T} H$, then $\mathcal{R}$ cannot be of type I. Instead of discussing all cases, we shall pick up the case of type II_∞ because the other cases are easier.

The case N={1} and the trivial χ and module: Let H_0 be a countable dense subgroup of the dual group $\hat{H}$, and consider an outer action of H_0 on $\mathcal{R}_0$ and let δ be the dual action restricted to H on the crossed product $\mathcal{R}_0 \rtimes H_0$ which is isomorphic to $\mathcal{R}_0$ itself by [27]. Let $p \in H_0 \longrightarrow a_p \in \mathcal{U}(\mathcal{R}_0 \rtimes H_0)$ be the associated unitary representation of H_0 so that $\rho_h(a_p) = \langle h,p \rangle\, a_p$, $p \in H_0$, $h \in H$. Now, set up the following system:

$$\mathcal{N} = \prod_{n \in \mathbb{Z}}^{\otimes} \mathcal{R}_n, \quad \mathcal{R}_n = \mathcal{R}_0 \rtimes H_0,\; n \in \mathbb{Z};$$

$$\alpha_h = \prod_{n\in\mathbb{Z}}^{\otimes} \rho_{h,n}, \qquad \rho_{h,n} = \rho_h;$$

$$\beta_p = \mathrm{id} \otimes \prod_{n\geqslant 1}^{\otimes} \mathrm{Ad}(a_p)_n;$$

$$v_p = \Big(\prod_{n\leqslant -1}^{\otimes} 1_n\Big)\otimes a_p \otimes \Big(\prod_{n\geqslant 1}^{\otimes} 1_n\Big).\ 1_n = 1;$$

$$\sigma = \text{shift to the left on } \mathfrak{N}.$$

It then follows that $(h,p) \in H \rtimes H_0 \longrightarrow \alpha_h \beta_p$ is an outer action of $H \times H_0$ and

$$\sigma\alpha_h\sigma^{-1} = \alpha_h;\ \sigma\beta_p\sigma^{-1} = \mathrm{Ad}(v_p)\beta_p;\ \alpha_h(v_p) = \langle h,p\rangle v_p.$$

Now, let φ be a bicharacter on $H \times K$. Then φ is of the form: $\varphi(h,\gamma) = \langle h, \Sigma(\gamma)\rangle$ where Σ is a homomorphism of K into $\hat{H}$. By the cohomology Lemma, Σ is equivalent to a homomorphism: $\gamma\in K \longrightarrow q(\gamma) \in H_0$. We then define an action κ of $H \times K$ on N by:

$$\kappa_{h,\gamma} = \alpha_h \beta_{q(\gamma)}, \qquad h\in H,\ \gamma\in K,$$

and we set $\theta_x = \sigma$ for all $x\in X = K^{(0)}$. We then have:

$$\theta_y\ \kappa_{1,\gamma}\theta_x^{-1} = \mathrm{Ad}(v_{q(\gamma)})\ \kappa_{1,\gamma},$$
$$\theta_x\kappa_{h,x}\theta_x^{-1} = \kappa_{h,x},$$

and $\gamma\in K \longrightarrow v_{q(\gamma)}$ is a cocycle since $p\in H_0 \longrightarrow v_p$ is a cocycle for β. Now we have

$$\kappa_{h,y}(v_{q(\gamma)}) = \alpha_h(v_{q(\gamma)}) = \langle h,q(\gamma)\rangle v_{q(\gamma)}.$$

We now want to show that κ is cocycle conjugate to the action: $(h,\gamma)\in H \times K \longrightarrow \alpha_g \otimes \mathrm{id}$ on $N\otimes N$. By Ocneanu's theorem, [18], the action of $H \times H_0$ given by α and β is cocyle conjugate to the action: $(h,p)\in H \times H_0 \longrightarrow \alpha_h\otimes\beta_p \in \mathrm{Aut}(\mathfrak{N}\bar{\otimes}\mathfrak{N})$. This means that κ is cocycle conjugate to the action: $(h,\gamma)\in H \times K \longrightarrow \alpha_h \otimes \beta_{q(\gamma)}$. Since Int(N) is a dense normal Borel subgroup of Aut(C), the Cohomology Lemma imples that there exist Borel functions: $x \in X$ $\sigma_x \in \mathrm{Aut}(N)$ and $\gamma \in K \longrightarrow u_\gamma \in \mathcal{U}(\mathcal{N})$ such that

$$\sigma_y \beta_{q(\gamma)} \sigma_x^{-1} = \mathrm{Ad}(u_\gamma).$$

Since the second cohomology of K in $\mathbb{T}$ vanishes, u_γ can be readjusted so that it is a homomorphism of K into $\mathcal{U}(\mathcal{N})$. Hence, the action κ is cocycle conjugate to the action $\bar{\alpha}$: $(h,\gamma)\in H \times K \longrightarrow \alpha_h \otimes \mathrm{id}$ on $\mathcal{N} \bar{\otimes} \mathcal{N}$.

Now, this $\bar{\alpha}$ has the following property: For any bicharacter φ: $(h,\gamma)\in H \times K \longrightarrow \mathbb{T}$, there exists a Borel map: $x \in X \longrightarrow \theta_x \in \mathrm{Aut}(\mathcal{R})$, where $\mathcal{R} = \mathcal{N} \bar{\otimes} \mathcal{N}$, such that

$$\theta_y \theta_x^{-1} = \mathrm{Ad}\, u_\gamma, (\gamma\colon x \longrightarrow y,\ \gamma \longrightarrow u_\gamma \text{ cocycle});$$
$$\theta_x \bar{\alpha}_h \theta_x^{-1} = \mathrm{Ad}(v_{h,x}) \bar{\alpha}_h,\ (h \longrightarrow v_{h,x} \text{ cocycle});$$
$$u_\gamma v_{h,x} = \varphi(h,\gamma) v_{h,y} \bar{\alpha}_h(u_\gamma),\ \gamma\colon x \longrightarrow y.$$

The case $N \neq \{1\}$ but the trivial module: We want a model action with arbitrary χ. To this end, we apply the above construction for H/N to obtain the action $\bar{\alpha}$ of $H/H \times K$. Next, we observe, based on the splitting $\mathfrak{G} \cong H \times K$, that

$$\Lambda(\mathfrak{G}, \mathcal{N}, \mathbb{T}) = H^1(K, \hat{N}) \otimes \Lambda(H, N, \mathbb{T}),$$

where $\mathcal{N} = N \times X$. Thus, $\chi \in \Lambda(\mathfrak{G}, \mathcal{N}, \mathbb{T})$, has two parts $\Sigma\colon \gamma \in K \longrightarrow \Sigma(\gamma) \in \hat{N}$ and $\chi \in \Lambda(H, N, \mathbb{T})$. Let $\chi = [\lambda, \mu]$. Fix an outer action σ of H on $\mathcal{R}_0$. Consider the twisted crossed product $\mathcal{R}_0 \rtimes_{\sigma,\mu} N$ and

let $\{u(k)\colon k \in N\}$ be the associated μ-representation of N into

$\mathcal{P} = \mathcal{R}_0 \rtimes_{\sigma,\mu} N$. Then define β of H on $\mathcal{P}$ by

$$\begin{cases} \beta_h(x)=\sigma_h(x), & x\in\mathcal{R}_0; \\ \beta_h(u(k))=\lambda(k,h)u(k), & k\in N, h\in H. \end{cases}$$

We then have $\chi_\beta = [\lambda,\mu]$. Now, we define

$$\begin{cases} \beta_\gamma(x)=x, x\in\mathcal{R}_0; \\ \beta_\gamma(u(k))=\langle k,\Sigma(\gamma)\rangle u(k), k\in N, \gamma\in K. \end{cases}$$

With this action β of $H \times K$, we have

$$\chi_\beta = \Sigma\otimes[\lambda,\mu]\in\Lambda(\mathcal{G},\mathcal{N},\mathbb{T}).$$

We now define

$$m = \bar{\alpha} \otimes \beta \text{ on } \mathcal{N} \bar{\otimes} \mathcal{P} \cong \mathcal{N}.$$

The general case: We keep the above m. Let δ be a module of $H \times K$. Let $\{\theta_t\}$ be the one parameter automorphism group of $\mathcal{R}_{0.1}=\mathcal{R}_0 \bar{\otimes} \mathcal{L}(\ell^2)$ such that $\mathrm{mod}(\theta_t)=e^{-t}$, which exists by the structure theorem for the Krieger factor of type III_1. The model action $\bar{m}$ is then defined by:

$$\bar{m}_{(h,\gamma)}=m_{(h,\gamma)}\otimes\theta_{-\log\delta(h,\gamma)}, \quad (h,\gamma)\in H \times K.$$

This model action $\bar{m}$ has the property that: i) χ_m is precisely the given χ; ii) the module of m is also the given δ; and, iii) for any bicharacter φ on $H \times K$ which vanishes on $N \times K$, there exists Borel map: $x\in X \longrightarrow \theta_x\in \mathrm{Aut}(\mathcal{N}\bar{\otimes}\mathcal{P}\bar{\otimes}\mathcal{R}_{0,1})$, $\gamma\in K \longrightarrow u_\gamma\in\mathcal{U}$, and $(h,x)\in H \times X \longrightarrow u_{h,x}\in\mathcal{U}$ such that

$$\begin{aligned} \theta\bar{m}_{h,x}\theta_x^{-1} &= \mathrm{Ad}(u_{h,x})\bar{m}_{h,x}; \ h\longrightarrow u_{h,x} \text{ cocycle;} \\ \theta_y\bar{m}_{1.\gamma}\theta_x^{-1} &= \mathrm{Ad}(u_\gamma)\bar{m}_{1.\gamma}: \ \gamma \longrightarrow u_\gamma \text{ cocycle;} \\ u_\gamma m_{1.\gamma}(u_{h,x}) &= \varphi(h,\gamma)u_{h,x}m_{h,x}(u_\gamma), \end{aligned}$$

where γ: $x \longrightarrow y$.

With this model action $\bar{m}$, we can prove the main theorem, Theorem 2. We refer the detail to the original paper [15].

REFERENCES

1. H. Araki and E.J. Woods, A classification of factors, Publ. Res. Inst. Math. Sci., **3**(1968), 51-130.

2. A. Connes, Une classification des facteurs de type III, Ann. Éc. Norm. Sup., **6**(1973), 133-252.

3. ___________, Outer conjugacy classes of automorphisms of factors, Ann. Éc. Norm. Sup., **8**(1975), 383-419.

4. ___________, Classification of injective factors, Cases II_1, II_∞, III_λ, $\lambda \neq 1$, Ann. Math., **104**(1976), 73-115.

5. ___________, Periodic automorphisms of the hyperfinite factor of type II_1, Acta. Sci. Math., **39**(1977), 39-66.

6. A. Connes, J. Feldman and B. Weiss, Amenable equivalence relations are hyperfinite, Ergodic Theory and Dynam. Sys., **1**(1981), 431-450.

7. A. Connes and W. Krieger, Measure space automorphisms, the normalizer of their full groups and approximate finiteness, J. Functional Analysis, **29**(1977), 336.

8. A. Connes and M. Takesaki, The flow of weights on factors of type III, Tôhoku Math. J., **29**(1977), 453-575; Errata.

9. H.A. Dye, On groups of measure preserving transformations, I, Amer. J. Math., **81**(1959), 119-159; II, ibid., **85**(1963), 551-576.

10. J. Feldman, and D. Lind, Hyperfiniteness and the Halmos-Rohlin theorem for non-singular abelian actions, Proc. Amer. Math. Soc., **55**(1976), 339-344.

11. J. Feldman and C. C. Moore, Ergodic equivalence relations,

cohomology, and von Neumann algebras, I, Trans. Amer. Math. Soc. **234**(1977), 289-324; II, ibid, 325-359.

12. J. Hakeda and J. Tomiyama, On some extension properties of von Newmann algebras, Tōhoku Math. J., **19**(1967), 315-323.

13. V. Jones, Actions of finite groups on the hyperfinite type II_1 factor, Mem. Amer. Math. Soc., **237**(1980).

14. __________, Prime actions of compact abelian groups on the hyperfinite II_1 factor, to appear.

15. V. Jones and M. Takesaki, Actions of compact abelian groups on semi-finite injective factors, Acta. Math., 153(1984), 213-258.

16. W. Krieger, On ergodic flows and the isomorphism of factors, Math. Ann., **224**(1976), 19-70.

17. G. W. Mackey, Ergodic theory and virtual groups, Math. Ann., **166**(1966), 187-207.

18. A. Ocneanu, Actions of discrete amenable groups on factors, to appear.

19. D. Ornstein and B. Weiss, Ergodic theory of amenable group actions, I. The Rohlin lemma, Bull. Amer. Math. Soc., **2**(1980), 161.

20. J. Phillips, Automorphisms of full II_1-factors, with applications to factors of type III, Duke Math. J., **43**(1976), 375-385.

21. R. T. Powers, Representations of uniformly hyperfinite algebras and their associated von Newmann rings, Ann. Math., **86**(1967), 138-171.

22. J. T. Schwartz, Two finite, non-hyperfinite, non-isomorphic factors, Comm. Pure and Appl. Math., **16**(1963), 19-26.

23. C. Sutherland, Notes on orbit equivalence: Kreiger's theorem, Lecture Notes Ser. No. 23, Oslo (1976).

24. ___________, Cohomology and extensions of von Neumann algebras, I, Publ. RIMS., Kyoto Univ., **16**(1980), 105–134; II, ibid., 135–174).

25. C. Sutherland and M. Takesake, Actions of amenable groups and groupoids on semifinite injective von Neumann algebras, to appear.

26. M. Takesaki, Theory of operator algebras, I, Springer-Verlag, (1979).

27. ___________, Duality for crossed products and the structure of von Neumann algebras of type III, Acta. Math., **131**(1973), 249–310.

28. ___________, Structure of factors and automorphism groups, CBMS. Regional Conference Ser. Math., **51**(1983).

Ergodic Therory and the Automorphism Group of a G-structure[1]

by

Robert J. Zimmer

Deapartment of Mathematics
University of Chicago
Chicago, IL 60637

Dedicated to George W. Mackey

1 Research partially supported by NSF Grant DMS-8301882

Contents

1. Introduction

A fundamental problem in geometry is to understand the automorphism group of a geometric structure. In this paper, we discuss the contribution that ergodic theory can make in this direction. We shall see that one can obtain new basic results, some quite definitive, that have not been obtained by purely geometric methods.

Let M be a compact n-manifold, $G \subset GL(n,\mathbb{R})$ a real algebraic group with Lie algebra $\mathfrak{g}$, and $P \longrightarrow M$ a G-structure on M. Thus P is a reduction of the frame bundle of M to G. Let $\mathrm{Aut}(P) \subset \mathrm{Diff}(M)$ be the automorphism group of the G-structure, i.e., the diffeomorphisms of M preserving P. The study of Aut(P) naturally breaks up into studying $\mathrm{Aut}(P)_0$, the connected component of the identity, and the group of connected components. In this paper, we shall focus on $\mathrm{Aut}(P)_0$, providing essentially complete proofs of our main results. We shall indicate some results for actions of discrete groups, but shall refer the reader elsewhere for proofs.

Theorem 5.5 [15] Let H be a connected non-compact simple Lie group with Lie algebra $\mathfrak{h}$. Let M be a compact n-manifold with a G-structure, where G is a real algebraic group. Suppose H acts smoothly on M preserving the G-structure and a volume density. Then there is a Lie algebra embedding $\pi: \mathfrak{h} \longrightarrow \mathfrak{g} \subset \mathfrak{gl}(n,\mathbb{R})$ such that as a representation of $\mathfrak{h}$ on $\mathbb{R}^n$, π contains $\mathrm{ad}_{\mathfrak{h}}$ as a direct summand.

(This result is also true for structures of higher order.)

For an explicit G we can obtain sharper results. We recall that the isometry group of a pseudo-Riemannian manifold is a Lie group [2]. A Lorentz manifold is of course a manifold with an O(1,n-1)-structure.

Theorem 9.1 [15] Let M be a compact manifold with Lorentz metric ω, and let $\mathrm{Aut}(M,\omega)$ be the isometry group with respect to ω. Then either:

a) $\mathrm{Aut}(M,\omega)_0$ is locally isomorphic to $SL(2,\mathbb{R}) \times K$, where K is a compact Lie group; in fact, for the universal cover we have

$$\widetilde{\mathrm{Aut}(M,\omega)_0} \cong \widetilde{SL(2,\mathbb{R})} \times C \times \mathbb{R}^r$$

where C is compact and simply connected, and $\mathbb{R}^r$ acts on M via a quotient torus; or

b) The radical of $\mathrm{Aut}(M,\omega)_0$ is cocompact, and the nilradical is at most a 2-step nilpotent group.

We observe in the final section of this paper that for any compact Lie group K, there is a compact Lorentz manifold whose isometry group contains $PSL(2,\mathbb{R})\times K$ as a closed subgroup, and that for any Heisenberg group N, there is a compact Lorentz manifold and a free N-action preserving the Lorentz metric.

Most of the remainder of this paper is devoted to an exposition of the proof of the above theorems. We have assumed very little background in ergodic theory and have included material designed to make the techniques involved more easily accessible to geometers. We conclude this introduction with the statement of some results concerning actions of discrete groups. For proofs and further results in this direction, we refer the reader to [10]-[15].

The discrete groups with which we are concerned are lattice subgroups $\Gamma\subset H$, where H is a connected, simple, non-compact Lie group with finite center, i.e. Γ is discrete and H/Γ has a finite H-invariant measure. Many such groups have dense embeddings in compact Lie groups. If K is a compact Lie group, then for any real algebraic subgroup $G\subset GL(\dim K,\mathbb{R})$ there is a K-invariant G-structure on K, and hence a Γ-invariant G-structure on K for any homomorphism $\Gamma \longrightarrow K$. The next theorem asserts that for certain G-structures of finite type (in the sense of E. Cartan [2]), any Γ-action preserving a G-structure must be of a similar nature. (We recall that if $G = O(p,q)$, then G is of finite type.)

<u>**Theorem**</u> [11],[15] Let $\Gamma\subset H$ be as above and assume the real (i.e. split) rank of H is at least 2. Let M be a compact manifold with a G-structure of finite type where G is real algebraic group. Suppose

Γ acts on M, preserving the G-structure and a volume density. Then either:

a) There is a Lie algebra embedding $\mathfrak{h} \longrightarrow \mathfrak{g}$;

or b) There is a Γ-invariant Riemanninan metric on M.

Corollary [15] Let $m \geq 3$, and $\Gamma \subset SL(m,\mathbb{R})$ be a lattice. Let M be a compact manifold with dim $M < m$. Then any action of Γ on M preserving any G-structure of finite type and any volume density is an action by a finite quotient of Γ.

Similar results are true if $SL(m,\mathbb{R})$ is replaced by any simple Lie group of $\mathbb{R}$-rank at least 2.

2. Measurable reductions of principal bundles.

A basic ingredient in the proofs of the theorems in the introduction is a systematic study of the measure theoretic properties of actions on principal bundles, and in particular the study of invariant measurable reductions of the bundle to a closed subgroup. More precisely, let $P \longrightarrow M$ be a (smooth) principal G-bundle over a manifold M where G is a Lie group. Thus, G acts freely on the right on P and $P/G \cong M$. Let $G_1 \subset G$ be a closed subgroup. We recall that a smooth reduction of P to G_1 is a smooth section r of the bundle $P/G_1 \longrightarrow M$. Thus, r chooses, in a smoothly varying way, a single G_1-orbit in each fiber over M. Alternatively, let $p_1 : P \longrightarrow P/G_1$ be the natural map and let $Q = p_1^{-1}(r(M))$. Then $Q \longrightarrow M$ is a principal G_1-bundle, $Q \subset P$, and P can be recovered from Q by the standard construction of forming the associated bundle for the G_1 action on G given by translation. All of this is of course totally standard.

Now let μ be a measure on M. (Often this will be the measure defined by a volume density, but not always.) For a given μ, a μ-measurable function or μ-measurable section will always be taken to mean an equivalence class of functions or sections, two being identified if they agree μ-a.e. If μ is understood, we shall simply refer to a measurable function or section.

Definition 2.1 Let P, M, G, G_1 be as above. A (μ-)measurable reduction of P to G_1 is a (μ-)measurable section $\varphi: M \longrightarrow P/G_1$.

Thus, φ still chooses a G_1-orbit in each (more precisely, for μ-a.e.) fiber over M, but these are now only required to vary measurably. We remark that while smooth reductions may or may not exist, measurable reductions always exist, as one can measurably piece together local sections. By way of illustration, suppose $E \longrightarrow M$ is a vector bundle of rank n and $P \longrightarrow M$ is the corresponding principal $GL(n,\mathbb{R})$-bundle of frames. Riemannian metrics on E are in bijective correspondence with (smooth) reductions of P to $O(n,\mathbb{R})$. By a (μ-) measurable Riemanninan metric on E, we naturally mean a mesurable section B of $S^2(E^*) \longrightarrow M$ (where $S^2(E^*)$ is the bundle of symmetric bilinear forms on E) such that B(m) is an inner product on E_m for (almost) all $m \in M$. Then we have:

Proposition 2.2 Measurable Riemannian metrics on E are in natural bijective correspondence with measurable reductions of P to $O(n,\mathbb{R})$.

The correspondence, of course, assigns to each measurable Riemannian metric the corresponding set of orthonormal frames.

Suppose now that H acts (on the left) on P by principal bundle automorphisms. In other words, H acts smoothly on P, commuting with the G-action, and the induced action of H on M is smooth. If $G_1 \subset G$, then H acts on P/G_1 and hence we may speak of an H-invariant reduction of P to G_1. In the case of a measurable reduction $\varphi: M \longrightarrow G/G_1$, by ($\mu$-) invariance we shall mean that for each $h \in H$, $\varphi(hm) = h\varphi(m)$, for a.e. $m \in M$, where μ is understood to be quasi-invariant under H, i.e. the null sets are preserved by the H-action. (At times, this a.e. condition causes some technical problems, but for the most part these are routine and can be safely ignored. In any case, we shall systematically ignore them in this paper.) For example, if H acts by vector bundle automorphisms on $E \longrightarrow M$, and hence by principal bundle automorphisms of the frame bundle $P \longrightarrow M$, then of course H-invariant measurable reductions of P to $O(n,\mathbb{R})$ simply correspond to H-invariant measurable Riemannian metrics on E.

In considering such measure theoretic properties of actions of H, it is useful to measurably trivialize the bundle $P \longrightarrow M$. Namely, if $s:M \longrightarrow P$ is any measurable section (as observed above, these always exist), then the map $\Phi:M\times G \longrightarrow P$ given by $\Phi(m,g) = s(m)g$ is a measurable bijection such that the bundle map $P \longrightarrow M$ corresponds to projection of $M\times G$ on the first factor. The corresponding action of G on $M\times G$ is of course right translation on the second factor, and if H acts by principal bundle automorphisms of P, the action of H on $M\times G$ is given by $h(m,g) = (hm,\alpha(h,m)g)$ where $\alpha(h,m)\in G$. The map $\alpha:H\times M \longrightarrow G$ is a measurable cocycle, namely:

Definition 2.3 Let H act on M and suppose μ is an H-quasi-invariant measure on M. A function $\alpha:H\times M \longrightarrow G$ is called a (μ-) cocycle if for each $h_1,h_2\in H$, $\alpha(h_1h_2,m) = \alpha(h_1,h_2m)\alpha(h_2,m)$ for a.e. $m\in M$.

(We remark that if we fix a Borel section of $P \longrightarrow M$, then we obtain a Borel map $\alpha:H\times M \longrightarrow G$ which will satisfy the cocycle identity at all points, and hence be a (μ-) cocycle for any quasi-invariant μ.)

Given any two measurable sections $s_1,s_2:M \longrightarrow P$, the corresponding cocycles α_1,α_2 are related as follows: there is a measurable $\varphi:M \longrightarrow G$ such that for each $h\in H$, $\alpha_2(h,m) = \varphi(hm)^{-1}\alpha_1(h,m)\varphi(m)$ for a.e. $m\in M$. (Here φ is defined by the equation $s_1(m) = s_2(m)\varphi(m)$.) Two cocycles α_1, α_2 satisfying such a relation for a measurable φ are called (μ-) equivalent and we write $\alpha_1 \sim \alpha_2$. Thus, the action of H on P naturally defines an equivalence class of cocycles $H\times M \longrightarrow G$.

Suppose now that $G_1\subset G$ is a closed subgroup. An H-invariant measurable section $M \longrightarrow P/G_1$ corresponds under Φ to a measurable map $\psi:M \longrightarrow G/G_1$ such for each $h\in H$, $\alpha(h,m)\psi(m) = \psi(hm)$ for a.e. $m\in M$. Quite generally, if $\alpha:H\times M \longrightarrow G$ is a cocycle and X is a G-space, we shall use the following terminology:

Definition 2.4 A measurable function $\psi:M \longrightarrow X$ is called

α-invariant if for each $h \in H$, $\alpha(h,m)\psi(m) = \psi(hm)$ a.e.

Fix a Borel section q of the natural projection $p: G \longrightarrow G/G_1$. Then if ψ is α-invariant, we have $p(\alpha(h,m))q(\psi(m))) = p(q(\psi(hm)))$, or in other words, letting $\lambda = q \circ \psi$, $\lambda(hm)^{-1}\alpha(h,m)\lambda(m) \in G_1$. Hence α is equivalent to a cocycle taking values in G_1. This, and the same argument run in reverse, shows:

Proposition 2.4 If H acts by automophisms of the principal G-bundle P, leaves a measure μ on M quasi-invariant, and $G_1 \subset G$ is closed, then the following are equivalent:

i) There is an H-invariant measurable reduction of P to G_1;

ii) A cocycle $\alpha: H \times M \longrightarrow G$ corresponding to the action is equivalent to a cocycle taking all values in G_1;

iii) There is an α-invariant function $M \longrightarrow G/G_1$.

Thus, for example, there is a measurable invariant Riemannian metric if and only if α is equivalent to a cocycle into $O(n,\mathbb{R})$.

This proposition points out the interest of the following question: Given a cocycle α into G, for which closed subgroups G_1 is it true that $\alpha \sim \beta$ where $\beta(H \times M) \subset G_1$? We will examine this problem in subsequent sections, but we conclude this section by exhibiting one relation of the notion of measurable invariant reductions to more standard geometric concepts.

Theorem 2.5 Let M be a compact manifold with a (not necessarily smooth) probability measure μ, and a connection. Let H be a group that acts smoothly on M, preserving μ and the connection. If there is a (μ-) measurable invariant Riemannian metric on M (i.e. on the tangent bundle), then there is a smooth H-invariant Riemannian metric on M.

Proof Let $A(M)$ be the group of affine (i.e. connection preserving) transformations of M. Then $A(M)$ is a Lie group, and by passing to the closure of the image of H in $A(M)$, we may assume that $H \subset A(M)$ is closed and that we need to show that H is compact. We recall that a connection on M (i.e. on the frame bundle $P \longrightarrow M$) and a fixed choice of a frame for the Lie algebra $\mathfrak{gl}(n,\mathbb{R})$ together determine a global A(M)-invariant framing on P. Thus, A(M) can be identified with a subgroup of the isometry group Iso(P), where P has the metric given by making the above frame orthonormal at each $p \in P$. It is also easy to see that $A(M) \subset \text{Iso}(P)$ is closed and hence that $H \subset \text{Iso}(P)$ is closed. The theorem then follows immediately from the following two lemmas.

Lemma 2.6 There is an H-invariant probability measure on P.

Lemma 2.7 Let N be a Riemannian manifold with finitely many connected components, and $H \subset \text{Iso}(N)$ a closed subgroup. If H leaves a probability measure on N invariant, then H is compact.

We now proceed to the (easy) proofs of the lemmas.

Proof of lemma 2.6 Since there is an H-invariant measurable Riemannian metric on M, by Proposition 2.4 we can choose a measurable section of the frame bundle such that the corresponding cocycle $\alpha: H \times M \longrightarrow GL(n,\mathbb{R})$ satisfies $\alpha(h,m) \in O(n,\mathbb{R})$ (for each h and μ-a.e. $m \in M$). Let ν be the Haar measure on $O(n,\mathbb{R})$. Then $\mu \times \nu$ will be an H-invariant measure on $M \times GL(n,\mathbb{R})$, and hence there is an H-invariant probability measure on P.

Proof of lemma 2.7 We recall the basic fact that Iso(N) acts properly on N. I.e. if $A, B \subset N$ are compact, then $\{h \in \text{Iso}(N) \mid hA \cap B \neq \emptyset\}$ is precompact. (See [3] e.g.) Let μ be the H-invariant measure on N, and $A \subset N$ a compact set with $\mu(A) > 0$. Suppose H is not compact. Then there is $h_1 \in H$ such that $h_1 A \cap A = \emptyset$. We can then choose $h_2 \in H$ such that $h_2 A, h_1 A, A$ are mutually disjoint. Continuing inductively, we can find a sequence $h_i \in H$ such that $h_i A$ are mutually disjoint. Since $\mu(A) > 0$, and $\mu(h_i A) = \mu(A)$, this clearly contradicts the finiteness of μ.

We remark that it is not true in general that a diffeomorphism preserving a measure and a measurable Riemannian metric preserves a smooth metric. Other conditions (i.e. other than the connection preserving assumption) on a group action which yield this implication are implicit in the proof of the theorems on actions of lattices described in the introduction. Theorem 2.5 will itself be of use later in the proof of Theorem 9.1 on the structure of the automorphism group of a Lorentz manifold. Theorem 2.5 and its proof generalize to the case of a general G-structure of finite type [2].

Theorem 2.8 [11][15] Let M be a compact manifold with a probability measure and a G-structure of finite type. Suppose H acts on M so as to preserve the measure and the G-structure. For each r, let $J^r(M) \longrightarrow M$ be the r-th jet bundle of real valued functions on M. If there is an H-invariant measurable Riemannian metric on $J^r(M)$ for a suitably large r, then there is a smooth H-invariant Riemannian metric.

As we will not be using this result, we refer the reader to [11], [15] for its proof.

3. Background in ergodic theory and the cocycle reduction lemma

This section has two purposes. One is to introduce the condition of ergodicity of an action and to show how one can reduce certain questions to the ergodic case. The second is to prove, in the ergodic case, an extremely useful general criterion for showing that a cocycle is equivalent to one taking values in a proper subgroup. For a more thorough treatment of some of this material, we refer the reader to [8].

Throughout this section X will be a compact metrizable space and H will be locally compact (second countable) group acting continuously on X. Let M(X) be the space of probability measures on X, so that M(X) is a compact convex subset of $C(X)^*$ with the weak-*-topology. Thus, H acts on M(X) as well. We let $I(X) \subset M(X)$ be the space of H-invariant probability measures. We shall assume $I(X) \neq \emptyset$ (which is not a priori always the case). Then I(X) is itself a compact convex set.

Definition 3.1 A measure $\mu \in M(X)$ which is quasi-invariant under the H-action (e.g. $\mu \in I(X)$) is called ergodic (or the action of H on (X,μ) is called ergodic) if $A \subset X$ is an H-invariant Borel set implies $\mu(A) = 0$ or $\mu(X-A) = 0$.

Before indicating what this is good for, let us indicate how to obtain information about a general $\mu \in I(X)$ from ergodic measures. We let $\mathrm{Erg}(X) = \{\mu \in I(X) \mid \mu \text{ is ergodic}\}$.

Proposition 3.2 If $\mu \in I(X)$ is an extreme point of $I(X)$, then $\mu \in \mathrm{Erg}(X)$.

Proof If μ is not ergodic, let $A \subset X$ be invariant with $\mu(A)\mu(X-A) \neq 0$. Let $\mu_1(B) = \mu(A \cap B)/\mu(A)$, $\mu_2(B) = \mu((X-A)\cap B)/\mu(X-A)$. Then $\mu = \mu(A)\mu_1 + \mu(X-A)\mu_2$.

In particular, elementary results of functional analysis then enable us to assert that $I(X) \neq \emptyset$ implies $\mathrm{Erg}(X) \neq \emptyset$. Moreover, by standard representation theorems for compact convex sets (see [5], e.g.), any element in a compact convex set is the "barycenter" of a measure supported on the extreme points. In our context, this means that there is a separable metrizable space E and an injective Borel function $E \longrightarrow \mathrm{Erg}(X)$, $t \longrightarrow \mu_t$, such that for any $m \in I(X)$ there is a probability measure ν on E such that $\mu = \int \mu_t d\nu(t)$, where the latter means that for any $f \in C(X)$, $\mu(f) = \int_{t \in E} \mu_t(f) d\nu(t)$. (Actually, we really have $E = \mathrm{Erg}(X)$ and $t \longrightarrow \mu_t$ is the identity, but then the notation becomes a bit confusing.) Standard approximation arguments then show:

Corollary 3.3 With the above notation, for any Borel set $A \subset X$, we have

$$\mu(A) = \int_{t \in E} \mu_t(A) d\nu(t).$$

In particular, if $m(A) = 0$ for all $m \in Erg(X)$, then $\mu(A) = 0$ for all $\mu \in I(X)$.

This simple reduction to the ergodic case will be extremely useful. We remark that if μ is a smooth measure on a manifold, it would be desirable if each μ_t were a smooth measure on a submanifold. This, however, need not be the case. Thus, the above decomposition leads one to measures which may not be very well behaved from the viewpoint of standard differential topology.

Suppose now that H acts ergodically on (X, μ) and that $f: X \longrightarrow \mathbb{R}$ is an H-invariant Borel function. The measure $\nu = f_*\mu$ on $\mathbb{R}$ (defined by $(f_*\mu)(A) = \mu(f^{-1}(A))$ is a probability measure on $\mathbb{R}$, and the invariance of f combined with the ergodicity of μ implies that for any Borel set $A \subset \mathbb{R}$, $\nu(A) = 0$ or 1. It is a simple exercise to show that this implies that ν is supported on a point. In other words, for any H-invariant Borel f, there is a conull set (i.e. of null complement) $X_0 \subset X$ such that $f | X_0$ is constant. We then say that f is essentially constant. If one examines what properties of $\mathbb{R}$ are relevant here, one comes to the following definition.

Definition 3.4 If Y is a topological space, Y is called tame if Y has countable basis for its topology and Y is T_0. (We recall that T_0 is the assertion that $y, z \in Y$ implies the existence of an open set U such that either $y \in U$ or $z \in U$, but not both.)

We then have:

Proposition 3.5 With H, X as above, let $\mu \in M(X)$ be ergodic. Let $f: X \longrightarrow Y$ be an H-invariant Borel function where Y is tame. Then f is essentially constant.

The importance of tameness is that it arises naturally.

Definition 3.6 If a group G acts continuously on a separable metrizable space Z, the action is called tame if Z/G is tame with the quotient topology.

Some elementary point set topology then shows:

Proposition 3.7 Suppose G is a locally compact second countable group acting continuously on a separable metrizable space Z. If every orbit is locally closed, the action is tame.

(See [8, Chapter 2] for discussion. There "tame" is called "smooth". Here this would obviously be confusing.)

Corollary 3.8

a) If G is compact, every action is tame.

b) If G is real algebraic group acting regularly on a real algebraic variety, the action is tame.

We can now prove the following useful result.

Theorem 3.9 [8](Cocycle reduction lemma) Suppose H acts ergodically on (X,μ) and $\alpha:H\times X \longrightarrow G$ is a cocycle. Let Z be a space on which G acts continuously and tamely. Suppose there is an α-invariant measurable function $\varphi:X \longrightarrow Z$. Then $\alpha\sim\beta$ where $\beta(H\times X)\subset G_z$ where G_z is the stabilizer in G of a point $z\in Z$. Furthermore, for almost all $x\in X$, $\varphi(x) \in Gz$.

Proof We have $\alpha(h,m)\varphi(m) = \varphi(hm)$. Thus, if we let $\tilde{\varphi}:X \longrightarrow Z/G$ be the composition of φ with the natual quotient map, we have $\tilde{\varphi}(m) = \tilde{\varphi}(hm)$. Thus, $\tilde{\varphi}$ is H-invariant and Proposition 3.5 implies that (by passing to a conull set) $\tilde{\varphi}(X)$ is a point. In other words, there is a single G-orbit, say Gz, in Z such that $\varphi(X)\subset G\cdot z$. As a G-space, this orbit will be isomorphic to G/G_z, and hence φ can be considered as an α-invariant map $X \longrightarrow G/G_z$. The result then follows from Proposition 2.4.

We shall be using this result a number of times throughout the remainder of the paper. The following theorem is an illustration of how it is used. We shall not actually be using this so we only indicate the proof. However, similar ideas arise in section 8 below. We recall that a group is amenable if it has a fixed point for every linear action on a compact convex set in a locally convex linear space. A connected Lie group is amenable if and only if it has a solvable normal cocompact subgroup, or equivalently, if every simple Lie subgroup is compact. (See [8, Chapter 4].)

Theorem 3.10 [7][8] Suppose H is amenable and that $\alpha: H \times X \longrightarrow GL(n,\mathbb{R})$ is a cocycle, where H acts ergodically on X. Then there is an amenable algebraic subgroup $G \subset GL(n,\mathbb{R})$ and a cocycle $\beta \sim \alpha$ such that $\beta(H \times X) \subset G$.

Proof Let V be the variety of full flags in $\mathbb{R}^n$, so that $GL(n,\mathbb{R})$ acts naturally (and transitively) on V. Let F(X,M(V)) be the space of measurable functions. We can identify $F(X,M(V)) \subset L^\infty(X,C(V)^*) \cong L^1(X,C(V))^*$. Further, one readily checks that F(X,M(V)) is a compact convex set with the weak-*-topology. The group H acts on F(X,M(V)) by $(h^{-1} \cdot \varphi)(x) = \alpha(h,x)^{-1}\varphi(hx)$. Since H is amenable, there is a fixed point in F(X,M(V)). This is precisely the assertion that there is an α-invariant $\varphi: X \longrightarrow M(V)$. The theorem now follows from Theorem 3.9 and the following two (non-trivial) results. (See [8] for a proof of both.)

i) [7] The $GL(n,\mathbb{R})$ action on M(V) is tame.

ii) [4] The stabilizers in G of elements of M(V) are amenable algebraic groups.

4. The algebraic hull of a linear cocycle.

We now turn to the invariant whose computation will lead us to the main results.

Theorem 4.1 Let (X,μ) be a separable metrizable H-space with a quasi-invariant probability measure μ. Suppose the action is ergodic. Let $\alpha: H \times X \longrightarrow GL(n,\mathbb{R})$ be a cocycle. Then:

1) There is a real algebraic group $G \subset GL(n,\mathbb{R})$ such that α is equivalent to a cocycle taking values in G, but α is not equivalent to a cocycle taking values in a proper algebraic subgroup of G.

2) Any two algebraic subgroups of $GL(n,\mathbb{R})$ with property (1) are conjugate in $GL(n,\mathbb{R})$.

3) If $\alpha \sim \beta$, where $\beta(H \times X) \subset L$ and L is algebraic, then L contains a conjugate of G.

Proof

(1) follows from the descending chain condition on algebraic subgroups. This condition also implies (3) once (2) established. Thus, suppose $G_1, G_2 \subset GL(n,\mathbb{R})$ satisfy the conclusions of (1). Then by Proposition 2.4 there are α-invariant maps $\varphi_i : X \longrightarrow G/G_i$ where we have set $Gl(n,\mathbb{R}) = G$. Then $(\varphi_1, \varphi_2) : X \longrightarrow G/G_1 \times G/G_2$ is also α-invariant. Since G_i are algebraic, the G-action on $G/G_1 \times G/G_2$ is tame (Corollary 3.8), and by Theorem 3.9, α is equivalent to a cocycle into $L = G_y$ for some $y \in G/G_1 \times G/G_2$. We may replace L by a conjugate and deduce that α is equivalent to a cocycle into a group of the form $G_1 \cap gG_2g^{-1}$ for some $g \in G$. By the minimality property of G_1, $G_1 \subset gG_2g^{-1}$, and similarly, the minimality property of G_2 implies $G_2 \subset g^{-1}G_1g$. Thus, G_1 and G_2 are conjugate.

<u>Definition</u> <u>4.2</u> The conjugacy class of G in Theorem 4.1 (or sometimes by abuse of notation, G itself) is called the algebraic hull of α.

<u>Example</u> <u>4.2</u> If H is amenable, any $\alpha : H \times X \longrightarrow GL(n,\mathbb{R})$ has an amenable algebraic hull by Theorem 3.10.

The following is a straightforward exercise in the definitions.

<u>Proposition</u> <u>4.4</u> Let $V \subset \mathbb{R}^n$ be a linear subspace and $G_V = \{g \in GL(n,\mathbb{R}) \mid g(V) = V\}$. Let $p : G_V \longrightarrow GL(V)$ be the natural surjection. Suppose $\alpha : H \times X \longrightarrow G_V \subset GL(n,\mathbb{R})$ is a cocycle with algebraic hull $L \subset G_V$. Then the algebraic hull of $p \circ \alpha$ is the Zariski closure of $p(L)$ (in which $p(L)$ is of finite index.)

(We recall that if $\pi : G \longrightarrow H$ is an $\mathbb{R}$-regular homomorphism of complex algebraic groups defined over $\mathbb{R}$, then $\pi(G)$ is an algebraic $\mathbb{R}$-group, but that $\pi(G_{\mathbb{R}}) \subset (\pi(G))_{\mathbb{R}}$ may be of non-trivial finite index.)

The next two sections will contain a number of examples and applications.

5. Applications to simple Lie groups.

A basic result needed for the computation of the algebraic hull of a cocycle for an action of a simple Lie group is the Borel density theorem.

Theorem 5.1 (Borel)[1] Let G be a Zariski connected, non-compact, simple real algebraic group and $H \subset G$ an algebraic subgroup. Suppose there is a G-invariant probability measure on G/H. Then H = G.

(The name of the theorem comes from the equivalent assertion that any subgroup $\Gamma \subset G$ for which G/Γ has a finite G-invariant probability measure must be Zariski dense in G.)

The proof we give is due to Furstenberg. (See [8])

Proof We first make some observations about probability measures on $\mathbb{P}^{n-1}$. Let $\mu \in M(\mathbb{P}^{n-1})$ and $g_j \in SL(n,\mathbb{R})$. Suppose $g_j \longrightarrow \infty$ in $SL(n,\mathbb{R})$. Then by passing to a subsequence we have for $h_j = g_j/\|g_j\|$ that $h_j \longrightarrow g$ where g is an $n \times n$ real matrix with $\|g\| = 1$ and det g = 0. Let N = ker(g), R = image(g). By passing to a subsequence we can assume $h_j N \longrightarrow M \subset \mathbb{R}^n$. Thus we have for $x \in \mathbb{P}^{n-1}$ that

a) $x \in \mathbb{P}(N)$ implies distance$(g_j x,, M) \longrightarrow 0$;

b) $x \notin \mathbb{P}(N)$ implies $g_j x \longrightarrow y \in R$.

Write $\mu = \mu_1 + \mu_2$ where μ_1 is supported on $\mathbb{P}(N)$ and $\mu_2(\mathbb{P}(N)) = 0$. Then by passing to a subsequence we can assume $g_j \mu \longrightarrow \nu$, $g_j \mu_i \longrightarrow \nu_i$ and $\nu = \nu_1 + \nu_2$. From (a) we deduce that ν_1 is supported on $\mathbb{P}(M)$, and from (b), a straghtforward argument using the dominated convergence theorem shows ν_2 is supported on $\mathbb{P}(R)$.

(See [8, Chapter 3] for details.) Thus, from the case where $g_j\mu = \mu$, we deduce:

Lemma 5.2 If $G \subset SL(n,\mathbb{R})$ is a closed subgroup and G fixes an element of $M(\mathbb{P}^{n-1})$, then either:

i) G is compact; or

ii) $\mu(\mathbb{P}(V)) > 0$ for some linear subspace $V \subset \mathbb{R}^n$ with $0 < \dim V < n$.

We can now complete the proof of Theorem 5.1. By a theorem of Chevalley, there is rational representation $\pi: G \longrightarrow SL(m,\mathbb{R})$ for some m such that H is the stabilizer of a point in $\mathbb{P}^{m-1}$. If $H \neq G$, then of course ker π is finite. Thus $G/\ker \pi$ can be identified with a closed subgroup of $SL(m,\mathbb{R})$ and G/H with an orbit in $\mathbb{P}^{m-1}$. We may further assume that this orbit linearly spans $\mathbb{R}^m$. Therefore, there is a $G/\ker\pi$-invariant element $\mu \in M(\mathbb{P}^{m-1})$ supported on this orbit. Since G is not compact, by lemma 5.2 there is a proper subspace $V \subset \mathbb{R}^m$ with $\mu(\mathbb{P}(V)) > 0$. Choose such a V of minimal dimension. If $g \in G$ and $\mu(gV \cap V) \neq 0$, it follows that $gV = V$. Since μ is finite and G-invariant, it follows that a subgroup of G of finite index leaves V invariant. Since G is Zariski connected, G leaves V invariant. Further, $\mathbb{P}(V)$ must intersect the G-orbit supporting μ and hence must contain this orbit. This contradicts the assumption that this orbit spans $\mathbb{R}^m$, completing the proof.

Now let H be a connected simple Lie group, $\mathfrak{h}$ its Lie algebra and H^* the Zariski connected component of the identity in $\mathrm{Aut}(\mathfrak{h})$. We recall that Ad(H) is of finite index in H^*.

Corollary 5.3 Let H be a non-compact connected simple Lie group, (S,μ) an ergodic H-space, μ finite and invariant. Let $\alpha: H \times S \longrightarrow GL(\mathfrak{h})$ be the cocycle $\alpha(h,s) = \mathrm{Ad}(h)$. Then the algebraic hull of α is H^*.

Proof Let $G \subset H^*$ be the algebraic hull. Then (Prop. 2.4) there is an α-invariant map $\varphi: S \longrightarrow GL(\mathfrak{h})/G$. In this case, an α-invariant map is simply an H-map $\varphi: S \longrightarrow GL(\mathfrak{h})/G$. Let $\bar{\varphi}$ be the composition of φ with the natural map $GL(\mathfrak{h})/G \longrightarrow$

$H^*\backslash GL(\mathfrak{h})/G$. Then $\bar{\varphi}$ is H-invariant, and hence by 3.5, 3.8, $\varphi(S)$ is (a.e.) contained in a single H^*-orbit in $GL(\mathfrak{h})/G$. Thus, there is an H-map $\varphi:S \longrightarrow H^*/gGg^{-1}$ where $g \in GL(\mathfrak{h})$. Since S has a finite H-invariant measure, so does H^*/gGg^{-1}, and since $H^*/Ad(H)$ is finite, there is a finite H^*-invariant measure on H^*/gGg^{-1}. By theorem 5.1, $gGg^{-1} = H$, proving the corollary.

Corollary 5.3 has a very useful application to the stabilizers of H-actions.

Theorem 5.4 [9] Let H be a non-compact connected simple Lie group, and S an H-space with finite invariant measure. Then for almost all $s \in S$, the stabilizer $H_s \in \{H$, discrete subgroups of H$\}$. If the action is ergodic then we have either $H_s = H$ on a conull set, or H_s is discrete on a conull set.

Proof By Corollary 3.3, it suffices to show this for an ergodic measure on S. Let $\mathfrak{h}_s$ be the Lie algebra of H_s and $Gr(\mathfrak{h})$ the union of the Grassmann varieties of $\mathfrak{h}$. Then $s \longrightarrow \mathfrak{h}_s$ is a measurable map $S \longrightarrow Gr(\mathfrak{h})$, and this is clearly an H-map where H acts on $Gr(\mathfrak{h})$ via $Ad(H)$. By Theorem 3.9, the cocycle α of Corollary 5.3 is equivalent to a cocycle into an algebraic subgroup $L \subset GL(\mathfrak{h})$ where L is conjugate to the stabilizer in H^* of $\mathfrak{h}_s$ for a.e. $s \in S$. By Corollary 5.3, we have $L = H^*$. Thus, for a.e. $s \in S$, $\mathfrak{h}_s \subset \mathfrak{h}$ is an ideal, and this implies the assertion of the theorem.

We can now deduce our first basic result about automorphism groups of G-structures. For simplicity we assume we are dealing with first order structures, although the same proof works for higher order structures as well.

Theorem 5.5 [15] Let H be a connected non-compact simple Lie group, acting non-trivially on a compact n-manifold M so as to preserve a G-structure. We assume that G is a real algebraic group, and that it defines a volume density on M. Then there is a Lie algebra embedding $\pi:\mathfrak{h} \longrightarrow \mathfrak{g} \subset \mathfrak{gl}(n,\mathbb{R})$ such that the representation

π of $\mathfrak{h}$ on $\mathbb{R}^n$ contains $ad_{\mathfrak{h}}$ as a direct summand.

Proof By Theorem 5.4 and Corollary 3.3 there is an ergodic H-invariant probability measure μ on M such that for μ-almost all $m \in M$, H_m is discrete. Let $V_m \subset TM_m$ be the tangent space to the H-orbit through m. For any $m \in M$ with H_m discrete, the map $H \longrightarrow M$, $h \longrightarrow hm$, differentiates to yield an identification $f_m : \mathfrak{h} \longrightarrow V_m$, and the diagram

$$\begin{array}{ccc} \mathfrak{h} & \xrightarrow{Ad(h)} & \mathfrak{h} \\ {\scriptstyle f_m}\Big\downarrow & & \Big\downarrow{\scriptstyle f_{hm}} \\ V_m & \xrightarrow{dh_m} & V_{hm} \end{array}$$

commutes. Thus, with respect to μ, the derivative cocycle $H \times M \longrightarrow GL(n,\mathbb{R})$ is equivalent to a cocycle of the form

$$\alpha(h,m) = \left[\begin{array}{c|c} Ad(h) & * \\ \hline 0 & * \end{array}\right]$$

where we can identify $\mathfrak{h} \subset \mathbb{R}^n$. Since H preserves a G-structure, the algebraic hull of α is contained in G and the result follows from Theorem 4.1, Prop. 4.4 and Corollary 5.3.

To apply Theorem 5.5 to Lorentz structures, we need the following fact about Lie algebras.

Lemma 5.6 [15] Let $\mathfrak{h}$ be a simple real Lie algebra of non-compact type. Suppose $\pi : \mathfrak{h} \longrightarrow \mathfrak{gl}(V)$ is a representation of $\mathfrak{h}$ such that

i) There is a non-degenerate symmetric bilinear form on V of signature (1,dim V−1) which is invariant under $\pi(\mathfrak{h})$,

ii) $ad_{\mathfrak{h}}$ is a subrepresentation of π.

Then $\mathfrak{h} = sl(2,\mathbb{R})$.

For the proof, which uses only standard Lie theoretic arguments, we refer the reader to [15, proof of Theorem 4.1]. From

5.5, 5.6 we then deduce:

Theorem 5.7 [15] Let H be a connected non-compact simple Lie group. If H acts non-trivially on an compact manifold, preserving a Lorentz structure, then H is locally isomorphic to $SL(2,\mathbb{R})$.

6. Applications to nilpotent Lie groups.

In this section we prove results for nilpotent groups which are of much the same nature as the results on simple groups in the preceding section. As with simple groups, the following is basic. (Cf. Theorem 5.1.)

Theorem 6.1 Let N be a connected, simply connected nilpotent Lie group. Suppose $H \subset N$ is a connected subgroup such that N/H has a finite N-invariant measure. Then H = N.

This result is well-known, but for completeness we include a proof.

Proof We argue by induction on dim N. We shall need the elementary fact that the intersection of connected subgroups of N is connected. This follows easily, e.g., from the fact that the exponential map for N is a diffeomorphism.

If N is abelian, $N \cong \mathbb{R}^n$, and the assertion is clear, For N non-abelian, we have that H[N,N] is a connected normal subgroup. Since we have a surjective N-map $N/H \longrightarrow N/H[N,N]$, there is an N-(and hence N/[N,N]-) invariant probability measure on $N/H[N,N] \cong (N/[N,N])/(H[N,N]/[N,N])$. Thus, from the validity of our assertion in the abelian case, we deduce that H[N,N] = N. Therefore, the inclusion map $[N,N] \longrightarrow N$ induces a bijection $[N,N]/H\cap[N,N] \longrightarrow N/H$. Since $H\cap[N,N]$ is connected, the inductive assumption implies $H\cap[N,N] = [N,N]$, i.e., $H \supset [N,N]$. Since H[N,N] = N, this implies H = N.

We recall that any connected simply connected nilpotent Lie

group has (uniquely) the structure of a real algebraic unipotent group.

Corollary 6.2 Let N be a connected simply connected nilpotent Lie group with Lie algebra $\mathfrak{n}$, S a compact N-space with an ergodic N-invariant probability measure. Let $\alpha: N \times S \longrightarrow GL(\mathfrak{n})$ be the cocycle $\alpha(h,s) = Ad(h)$. Then the algebraic hull of α is Ad(N).

Proof Let $H \subset Ad(N)$ be the algebraic hull. Since Ad(N) is unipotent, H is connected. Arguing exactly as in the proof of Corollary 5.3, we deduce that there is an N-map $\varphi: S \longrightarrow Ad(N)/Ad(N) \cap gHg^{-1}$ for some $g \in GL(\mathfrak{n})$. Hence, there is an N-invariant probability measure on $Ad(N)/Ad(N) \cap gHg^{-1}$. Since Ad(N) and gHg^{-1} are both connected unipotent groups, so is $Ad(N) \cap gHg^{-1}$. By Theorem 6.1, $Ad(N) \cap gHg^{-1} = Ad(N)$, so $gHg^{-1} \supset Ad(N)$. Thus we have $\dim H \geq \dim Ad(N)$ and $H \subset Ad(N)$, showing that $H = Ad(N)$.

As in the case of simple groups, we can use Corollary 6.2 to deduce certain results about stabilizers of N-actions.

Theorem 6.3 Let N be a connected simply connected nilpotent Lie group and X a compact N-space with an invariant probability measure m whose support is X. Assume the action is locally faithful (i.e. it has a discrete kernel.) Then there is a positive integer r and an ergodic N-invariant probability measure ν on X^r such that for ν-almost all $y \in X^r$, the stabilizer N_y is discrete. (Here N acts on X^r via the product action.)

We first prove the following.

Lemma 6.4

a) For any N-ergodic $\mu \in M(X)$, there is a normal subgroup $N_\mu \subset N$ such that for μ-almost all $x \in X$, $N_x^0 = N_\mu$ where N_x^0 is the connected component of the identity of the stabilizer N_x.

b) For m-almost all $x \in X$, N_x^0 is normal.

Proof By Corollary 3.3, it suffices to see (a). Let $L(N_x)$ be the Lie algebra of N_x. Then $\varphi(x) = L(N_x)$ defines a measurable map $\varphi:X \longrightarrow Gr(L(N))$, the latter being the union of the Grassmann varieties of the Lie algebra of N. Furthermore, φ is an N-map where N acts on $Gr(L(N))$ via $Ad(N)$. Thus, letting $p:Gr(L(N)) \longrightarrow Gr(L(N))/Ad(N)$, we have that $p\circ\varphi$ is an N-invariant map, and hence by 3.5, it is essentially constant. Therefore, there is an N-orbit in $Gr(L(N))$ such that μ-almost all $\varphi(x)$ lie in this orbit. We can then also consider φ as an N-map $(X,\mu) \longrightarrow N/N_1$ where N_1 is the stabilizer of a point in this distinguished orbit. Clearly, N_1 is algebraic (and unipotent) and hence connected. Since X has a finite N-invariant measure, so does N/N_1, so Theorem 6.1 implies $N_1 = N$. Thus, for μ-almost all $x\in X$, the stabilizer of $L(N_x)$ in N is N itself, and hence N_x^0 is normal. Therefore the map $\varphi:X \longrightarrow Gr(L(N))$ is actually N-invariant, and by ergodicity it is constant on a conull set. This proves the lemma.

Consider now $L = \bigcap_{\mu\in Erg(X)} N_\mu$. This is a connected normal subgroup and by Corollary 3.3, any $g\in L$ fixes m-almost all points in X. Since $supp(m) = X$ and the action is locally faithful, it follows that $L = \{e\}$. By the descending chain condition on algebraic subgroups, there is a finite set of N-ergodic measures on X, say $\mu_1,\ldots,\mu_r$, such that $\bigcap_{i=1}^{r} N_{\mu_i} = \{e\}$. The measure $\mu_1\times\ldots\times\mu_r$ on X^r is N-invariant for the product action, and for a.e. $y\in X^r$ we have $N_y^0 = \{e\}$. The existence of an ergodic N-invariant measure on X^r with this property then follows from 3.3. This completes the proof of Theorem 6.3.

Theorem 6.5 Let N be a nilpotent Lie group with a locally faithful action on a compact manifold M, preserving a Lorentz metric. Then N is at most a 2-step nilpotent group, i.e., $Ad(N)$ is abelian.

Proof We can clearly assume N is simply connected. Let ν be the measure on M^r given by Theorem 6.3. Then the algebraic hull L of the derivative cocycle $\alpha:N\times M^r \longrightarrow GL(nr,\mathbb{R})$ clearly satisfies

$L \subset \prod_{i=1}^{r} O(1,n-1)$ since the N-action preserves a Lorentz structure on M. On the other hand, the N action has discrete stabilizers for ν-a.e. $y \in M^r$, and hence, as in the proof of Theorem 5.5, we can write $\mathfrak{n} \subset \mathbb{R}^{nr}$ in such a way that the cocycle is equivalent to $\beta: N \times M^r \longrightarrow GL(nr,\mathbb{R})$ where $\beta(g,y)|\mathfrak{n} = Ad_N(g)$. Replacing L by a conjugate, we can therefore assume L leaves $\mathfrak{n}$ invariant. In fact, letting $p: L \longrightarrow GL(\mathfrak{n})$ be the restriction map, we have $p(L) \subset Ad_N(N)$. By Theorem 6.1 and Prop. 4.4, we deduce that $p(L) = Ad(N)$. Summarizing, we have shown that there is an algebraic subgroup $L \subset \prod_{i=1}^{r} O(1,n-1)$ and a rational surjective homomorphism $p: L \longrightarrow Ad(N)$. It follows, if we let V be the unipotent radical of L, that $p(V) = Ad(N)$. However, any unipotent subgroup of $O(1,n-1)$ is abelian and hence the same is true for any unipotent subgroup of $\pi O(1,n-1)$. Thus, $Ad(N)$ is abelian, and this completes the proof.

The same proof yields further results on nilpotent Lie groups acting so as to preserve other G-structures. For example:

Theorem 6.6 Let N be a connected nilpotent Lie group acting locally faithfully on a compact n-manifold M, preserving a volume density. Then $l(N) \leq n$ where $l(N)$ is the length of the central series.

7. A digression in ergodic theory.

In sections 5,6 we examined the possible simple and nilpotent automorphism groups of Lorentz manifolds. To deal with the general case, we will need some further ergodic theoretic information of a general nature. The first result we will need presents a general criterion for the existence of a measurable invariant metric.

Definition 7.1 Let H be a locally compact group acting on a compact space X with an ergodic invariant probability measure μ. Let $\alpha: H \times X \longrightarrow G$ be a cocycle where G is locally compact. Then α is called strongly unbounded if for every $A \subset X$ with $\mu(A) > 0$ and a.e. $x \in A$, there is a sequence $h_n \in H$ such that

i) $h_n x \in A$

and ii) $\alpha(h_n, x) \longrightarrow \infty$ in G.

Theorem 7.2 (K. Schmidt) With H, X, G, α as above, the following are equivalent:

i) α is not strongly unbounded;

ii) $\alpha \sim \beta$ where $\beta(H \times X) \subset K$ for some compact subgroup $K \subset G$.

Proof Since any measurable function $X \longrightarrow G$ is bounded on a set of positive measure, it is clear that (ii) implies (i). To see the converse, we will need the following technical fact which is useful for reducing questions about an ergodic H-space X to a subset of X.

Lemma 7.3 Let X be an ergodic H-space and $A \subset X$ of positive measure.

a) Let Y be a tame space (Def. 3.4) and $f: A \longrightarrow Y$ a measurable function such that for each $h \in H$ and (a.e.) $x \in A$ with $hx \in A$, we have $f(hx) = f(x)$. Then f is constant on (a conull set in) A.

b) If $\alpha: H \times X \longrightarrow G$ is a cocycle, Z is a G-space, and $\varphi: A \longrightarrow Z$ is a measurable map such that $x \in A$, $h \in H$ and $hx \in A$ implies $\alpha(h,x)\varphi(x) = \varphi(hx)$, then φ extends to an α-invariant $\tilde{\varphi}: X \longrightarrow Z$.

c) If $\alpha: H \times X \longrightarrow G$ is a cocycle and there is a map $\varphi: A \longrightarrow G$ such that for $x \in A$, $h \in H$, $hx \in A$, we have $\varphi(hx)^{-1}\alpha(h,x)\varphi(x) \in G_1$ where $G_1 \subset G$ is closed subgroup, then α is equivalent to a cocycle into G_1.

Proof (We shall ignore certain technicalities involving null sets.) We can choose a measurable function $\lambda: X \longrightarrow A$ such that x and $\lambda(x)$ are in the same H-orbit for (almost) all $x \in X$. Write $x = h_x \lambda(x)$ where $x \longrightarrow h_x$ is a measurable map $X \longrightarrow H$. We can assume $h_x = e$ if $x \in A$. To see (a), define $\tilde{f}(x) = f(\lambda(x))$. Then $\tilde{f}: X \longrightarrow Y$ is H-invariant, and therefore constant on a conull set in X. Since $\tilde{f}|A = f$, (a) follows. Similarly, to see (b), for $x \in X$ let $\tilde{\varphi}(x) = \alpha(h_x, \lambda(x))\varphi(\lambda x)$. It is routine to verify that $\tilde{\varphi}$ is α-invariant. Finally, to see (c), we remark that $\alpha(h,x)\bar{\varphi}(x) = \bar{\varphi}(hx)$ where $\bar{\varphi}: A \longrightarrow G/G_1$ is the composition of φ with the natural projection $G \longrightarrow$

G/G_1. Assertion (c) then follows from (b) and Proposition 2.4.

We now return to the proof of Theorem 7.2. Suppose α is not strongly unbounded. Then there is a set A of positive measure and a set $B\subset A$ of positive measure such that for each $x\in B$, $C_x = \{\alpha(h,x) \mid h\in H \text{ such that } hx\in A\}$ has compact closure $\bar{C}_x\subset G$. If $x\in B$ and $a\in H$ with $ax\in B$, we clearly have $C_x = C_{ax}\alpha(a,x)$. Let S be the space of compact subsets of G. (We recall that S is a separable metrizable space where $K_n \longrightarrow K$ if and only if for any open neighborhood $U\supset K$ we have $K_n\subset U$ for all large n.) G acts on S via the natural left action: $g\cdot K = Kg^{-1}$. The map $\Phi:B \longrightarrow S$, $\Phi(x) = \bar{C}_x$ is measurable and satisfies $\alpha(a,x)\Phi(x) = \Phi(ax)$ if $x,ax \in B$. Thus by Lemma 7.3(b), Φ extends to an α-invariant map $\Phi:X \longrightarrow S$. It is easy to see that the G-action on S is tame and that the stabilizers are compact. It follows from Theorem 3.9 that α is equivalent to a cocycle into a compact subgroup.

The second general result of ergodic theory that we shall need is a sharpening of the integral decomposition of Corollary 3.3. If H acts on a compact metrizable space X, and $\mu,\nu \in M(X)$, $\mu\neq\nu$, are H-invariant and ergodic, then $\mu\perp\nu$. I.e. there are measurable sets $A,B\subset X$ such that $A\cap B = \emptyset$, $\mu(A) = 1$, $\nu(B) = 1$. (To see this recall by general measure theory that we can write $X = A\cup B\cup C$ where A,B,C are mutually disjoint, $\nu(A) = 0$, $\mu(B) = 0$, and μ and ν are mutually absolutely continuous on C. By invariance of μ and ν it follows that $hC = C$ (up to both μ-null sets and ν-null sets) and by ergodicity we deduce that either μ and ν are mutually absolutely continuous or they are disjoint. If they are absolutely continuous, $d\mu/d\nu$ will be H-invariant, hence constant, and since both are probability measures, $\mu = \nu$. Thus, if $\mu\in M(X)$ is any H-invariant measure, and we write $\mu(A) = \int_{t\in E} \mu_t(A)d\nu(t)$ where $\mu_t \in \text{Erg}(X)$ as in Corollary 3.3, then for any $t_1,t_2\in E$, $t_1\neq t_2$, we have $\mu_{t_1}\perp\mu_{t_2}$. The result we shall need is that the measurable decomposition of μ can be carried out for X as well. Namely:

Theorem 7.4 (Ergodic decomposition) There is a separable metrizable space E and an injective Borel map $E \longrightarrow Erg(X)$, $t \longrightarrow \mu_t$, such that for any H-invariant $\mu \in M(X)$ there are:

i) a measure ν on E such that

$$\mu(\Lambda) = \int_{t \in E} \mu_t(\Lambda) d\nu(t); \qquad \text{(this is}$$

Corollary 3.3);

and ii) a measurable map $p: X \longrightarrow E$ such that for ν-almost $t \in E$,

a) μ_t is supported on $E_t = p^{-1}(t)$,

and b) E_t is H-invariant.

This decomposition is thus one of measure theory while that in 3.3 is basically one of abstract functional analysis. For a proof, see [6]. The spaces (E_t, μ_t) are called the H-ergodic components of (X, μ).

8. Measures on the sphere under the conformal group.

For our application of the results of section 7 we will need some information on the behavior of elements of $M(S^{n-1})$ under the natural conformal action of $O(1,n)$ on S^{n-1}. This is in fact a special case of the general problem of understanding the behavior of elements of $M(G/P)$ under the natural G-action, where G is a real algebraic group and P is a cocompact algebraic subgroup. We have indicated some results about the general situation in the proof of Theorem 3.10. See [8, Chapter 3] for a more complete discussion. In our special case the result we need is the following.

Theorem 8.1 Let $g_i \in O(1,n)$ be a sequence with $g_i \longrightarrow \infty$ in $O(1,n)$. Then there is a subsequence g_{i_j} and a uniquely determined (by the subsequence) pair of points $\{x,y\} \subset S^{n-1}$ (not necessarily distinct) such that for any $\mu \in M(S^{n-1})$, $g_{i_j}\mu \longrightarrow \nu \in M(S^{n-1})$ where $\nu(\{x,y\})$

$= 1.$

Proof We can write $O(1,n) = KAK$ where $K = O(n,\mathbb{R})$ and $A \cong \mathbb{R}^+$. The action of K on S^{n-1} is the standard one and the action of A is given as follows. Let $x_0, y_0 \in S^{n-1}$ be opposite points, and $p: S^{n-1} - \{x_0\} \longrightarrow \mathbb{R}^{n-1}$ be stereographic projection on the plane tangent to y_0. Let $t \in \mathbb{R}^+ (\cong A)$ and M_t be multiplication on $\mathbb{R}^{n-1}$ by t. Then t acts on S^{n-1} by leaving x_0 fixed and acting on $S^{n-1} - \{x_0\}$ by $p^{-1} M_t p$. We can write $g_i = k_i a_i \tilde{k}_i$. If $g_i \longrightarrow \infty$, then $a_i \longrightarrow \infty$ and by passing to a subsequence g_{i_j}, we can assume $k_{i_j} \longrightarrow k \in K$, $\tilde{k}_{i_j} \longrightarrow \tilde{k} \in K$. If $x \in S^{n-1}$ and $x \neq \tilde{k}^{-1} y_0$, then $g_{i_j} x \longrightarrow k x_0$. On the other hand, by refining the subsequence further, we can assume $a_{i_j} \tilde{k}_{i_j} (\tilde{k}^{-1} y_0) \longrightarrow z_0 \in S^{n-1}$ for some z_0. Thus, $g_{i_j} \tilde{k}^{-1} y_0 \longrightarrow k z_0$. Then for any $\mu \in M(S^{n-1})$ we can show as in the proof of Theorem 5.1 (and the references listed there) that $g_{i_j} \mu \longrightarrow \mu$ where μ is supported on $\{k x_0, k z_0\}$.

Theorem 8.1 leads to the following result for $O(1,n)$-valued cocycles.

<u>**Lemma 8.2**</u> Let (X,μ) be a compact H-space (H any locally compact group) where $\mu \in M(X)$ is invariant and ergodic. Let $\alpha: H \times X \longrightarrow O(1,n)$ be a cocycle. Let $O(1,n)$ act on $M(S^{n-1})$ via the conformal action on S^{n-1}. Suppose there is an α-invariant function $\alpha: X \longrightarrow M(S^{n-1})$. Then exactly one of the following assertions is true.

a) α is equivalent to a cocycle into $O(n,\mathbb{R})$.

b) α is not equivalent to a cocycle into $O(n,\mathbb{R})$; there is a unique α-invariant function $\Phi: X \longrightarrow \{$2-point sets in $S^{n-1}\}$

c) α is not equivalent to a cocycle into $O(n,\mathbb{R})$; there is no α-invariant function $X \longrightarrow \{$2-point sets in $S^{n-1}\}$; there is a unique α-invariant function $X \longrightarrow S^{n-1}$.

Proof If (a) fails to hold then α is strongly unbounded by Theorem 7.2. We claim that for almost all $x \in X$, $\varphi(x)$ is supported on one or two points. Let $M_0 \subset M(S^{n-1})$ be the space of measures supported on one or two points. Then M_0 is a closed $O(1,n)$-invariant subset. Suppose there is a set $A \subset X$ of positive measure such that $\varphi(x) \notin M_0$ for all $x \in A$. Then there is an open set U, $M_0 \subset U \subset M(X)$ and a set $B \subset A$ of positive measure such that $x \in B$ implies $\varphi(x) \notin U$. However, if $x \in B$ and $h \in H$, $\alpha(h,x)\varphi(x) = \varphi(hx)$. Since α is strongly unbounded, for each $x \in B$ we can choose $h_n \in H$ such that $h_n x \in B$ and $\alpha(h_n,x) \longrightarrow \infty$. By theorem 8.1, by passing to a subsequence we can assume $\alpha(h_n,x)\varphi(x)$ converges to an element of M_0. Hence $\varphi(h_n x)$ converges to an element of M_0. Since $h_n x \in B$, this is impossible. Therefore $\varphi(x) \in M_0$ for (almost) all $x \in X$.

Now let S be the set of one or two point subsets of S^{n-1}. We can identify $S^{n-1} \subset S$ as a closed subset. Replacing $\varphi(x)$ by its support, we obtain an α-invariant map $\Phi: X \longrightarrow S$. Suppose there is an α-invariant map $\Psi: X \longrightarrow S-S^{n-1}$ and that α is strongly unbounded. It follows via an argument similar to that of the preceding paragraph that Ψ is then the unique α-invariant map $X \longrightarrow S-S^{n-1}$. Finally, if both (a), (b) fail, then $\Phi: X \longrightarrow S^{n-1}$, and again strong unboundedness combined with Theorem 8.1 shows that Φ is unique via the argument of the preceding paragraph. This verifies the lemma.

A basic hypothesis in Lemma 8.2 is the existence of an α-invariant map $X \longrightarrow M(S^{n-1})$. The space of such measurable functions, say $F(X,M(S^{n-1}))$ can be identified with a subset of $L^\infty(X,C(S^{n-1})^*) \cong L^1(X,C(S^{n-1}))^*$. It is not hard to see that $F(X,M(S^{n-1}))$ is a compact convex subset of $L^\infty(X,C(S^{n-1})^*)$ with the weak-*-topology. We recall that any solvable (or more generally, amenable) group has a fixed point for a linear action on a compact convex set (Kakutani-Markov fixed point theorem [8, Chapter 4]). Since the condition of being an α-invariant function is exactly the condition for being a fixed point for the corresponding H-action, the hypothesis on the existence of α-invariant function will always be satisfied if H is solvable. Combining these remarks, lemma 8.2 and the ergodic decomposition (Theorem 7.4), we arrive at the following.

<u>**Theorem 8.3**</u> Let H be a solvable group and (M,μ) a compact

H-space with an H-invariant probability measure. Suppose $\alpha: H \times X \longrightarrow O(1,n)$ is a cocycle. Then we can write (up to null sets) $M = M_1 \cup M_2 \cup M_3$ where M_i are mutually disjoint H-invariant measurable sets such that:

i) $\alpha | H \times M_1$ is equivalent to a cocycle into $O(n,\mathbb{R})$.

ii) For any H-invariant measurable set $A \subset M_2$ of positive measure, $\alpha | H \times A$ is not equivalent to a cocycle into $O(n,\mathbb{R})$; there is a unique $\alpha | H \times M_2$-invariant function $M_2 \longrightarrow S - S^{n-1}$ (where S is the space of one or two-point sets in S^{n-1}).

iii) For any H-invariant measurable set $A \subset M_3$ of positive measure, $\alpha | H \times A$ is not equivalent to a cocycle into $O(n,\mathbb{R})$, and there is no $\alpha | H \times A$-invariant function $A \longrightarrow S - S^{n-1}$; there is a unique $\alpha | H \times M_3$ invariant function $M_3 \longrightarrow S^{n-1}$.

Furthermore, the sets M_i are uniquely determined (up to null sets) by these properties.

9. Completion of the proof.

We can now prove our main theorem on the automorphism group of a compact Lorentz manifold.

Theorem 9.1 [15] Let M be a compact manifold with Lorentz metric ω and let $\mathrm{Aut}(M,\omega)$ be the isometry group with respect to ω. Then either:

a) $\mathrm{Aut}(M,\omega)_0$ is locally isomorphic to $SL(2,\mathbb{R}) \times K$, where K is a compact Lie group; in fact, for the universal cover we have $\widetilde{\mathrm{Aut}(M,\omega)}_0 \cong \widetilde{SL(2,\mathbb{R})} \times C \times \mathbb{R}^r$ where C is compact and simply connected and $\mathbb{R}^r$ acts on M via a quotient torus;

or b) The radical of $\mathrm{Aut}(M,\omega)_0$ is cocompact, and the nilradical is at most a 2-step nilpotent group.

In light of Theorems 5.7, 6.5, and the structure of Lie groups, it suffices to prove the following.

Lemma 9.2 Let G be a connected Lie group locally isomorphic to SL(2,ℝ), N a solvable Lie group on which G acts by automorphisms (perhaps trivially) and $H = G \ltimes N$. Suppose (M,ω) is a compact Lorentz manifold and $\pi: H \longrightarrow \mathrm{Aut}(M,\omega)$ is defined by a smooth locally faithful action, and $\pi(H)$ is closed. Then $\pi(N)$ is compact.

Proof Write $M = M_1 \cup M_2 \cup M_3$ for the N-action as in Theorem 8.3. If M_1 is not null, then by Theorem 2.5, $\pi(N)$ is compact. Thus, it suffices to see that M_2 and M_3 are null. If M_3 is not null, it is actually H-invariant by the uniqueness property of M_3 (since $N \subset H$ is normal). Furthermore, the uniqueness property of the $\alpha | N \times M_3$ invariant map $\varphi: M_3 \longrightarrow S^{n-1}$ implies that φ is actually $\alpha | H \times M_3$-invariant. By Proposition 2.4, this implies that for each H-ergodic component $Y \subset M_3$ (Theorem 7.4, ff.), the algebraic hull of $\alpha | H \times Y$ is contained in the stabilizer in $O(1,n)$ of a point in S^{n-1}. Hence this algebraic hull is amenable, i.e. every simple Lie subgroup is compact. If G does not fix almost every point in M_3, by Theorems 5.4, 7.4 there is an ergodic measure on M_3 such that for almost all points, the stabilizer in G is discrete. However, the proof of Theorem 5.5 then shows that the algebraic hull of $\alpha | G \times M_3$ (with respect to this ergodic measure) contains a group locally isomorphic to G. By amenability of this algebraic hull, this is impossible. Therefore, G must fix (almost) all points in M_3. But once again, this implies that for $m \in M_3$, $\alpha | G \times \{m\}$ is homomorphism of G into the stabilizer of a point in S^{n-1}. Thus $\alpha | G \times \{m\}$ is trivial. Summarizing, we have shown that if M_3 is not null, then for almost all $m \in M_3$, $gm = m$ for $g \in G$, and $dg_m = I$ for $g \in G$. However, since g preserves a Lorentz metric, this implies g acts by the identity on all of M. [3, Theorem I.3.2]. Thus, G acts trivially on M, which contradicts our assumption. Therefore, M_3 is null. A similar argument shows M_2 is null, and this completes the proof.

10. Examples.

Proposition 10.1 For any compact Lie group K, there is a compact Lorentz manifold (M,ω) such that $PSL(2,\mathbb{R}) \times K$ acts effectively on M, preserving ω.

Proof Let θ be the Killing form on $sl(2,\mathbb{R})$. If Γ is a discrete cocompact subgroup of $PSL(2,\mathbb{R})$, θ induces a $PSL(2,\mathbb{R})$ invariant Lorentz metric ω_1 on $M_1 = PSL(2,\mathbb{R})/\Gamma$. Let ω_2 be a K-invariant Riemannian metric on K. Then ω_1, ω_2 yield a $PSL(2,\mathbb{R})\times K$-invariant Lorentz structure on $M_1 \times K$.

Proposition 10.2 Let N be a Heisenberg group. Then there is a compact manifold M on which N acts freely and preserves a Lorentz metric.

Proof We prove this for the 3-dimensional Heisenberg group. The same proof works in general. Let $\mathfrak{n}$ be the Lie algebra of N, with basis $\{x,y,z\}$ such that $[x,y] = z$ and $\mathbb{R}z$ is the center. Let $\mathfrak{n}'$ be another copy of $\mathfrak{n}$, with similar basis $\{x',y',z'\}$. With respect to the basis $\{x,y,z,z',y',x'\}$ of $\mathfrak{n}\times\mathfrak{n}'$, the matrix

$$\begin{bmatrix} 1&0&0&0&0&0 \\ 0&1&0&0&0&0 \\ 0&0&0&1&0&0 \\ 0&0&1&0&0&0 \\ 0&0&0&0&1&0 \\ 0&0&0&0&0&1 \end{bmatrix}$$

defines an $Ad(N\times N)$-invariant symmetric bilinear form of signature (1,5). This induces an $N\times N$-invariant Lorentz structure on $(N\times N)/\Lambda$ where Λ is any discrete cocompact subgroup of $N\times N$. If $N\times\{e\}$ intersets no conjugate of Λ, it acts freely on $(N\times N)/\Lambda$, and this would prove the proposition. To construct such a Λ, let $\sigma \in Gal(\mathbb{Q}(\sqrt{2})/\mathbb{Q})$, $\sigma \neq e$. Then $\Lambda = \{(\gamma,\sigma(\gamma)) \in N\times N \mid \gamma \in N\cap GL(3,\mathbb{Z}[\sqrt{2}])\}$ suffices, where $N\subset GL(3,\mathbb{R})$ is the standard embedding.

REFERENCES

1. A. Borel, Density properties for certain subgroups of semisimple Lie groups without compact factors, Annals of Math., 72(1960), 179-188.
2. S. Kobayashi, Transformation groups in differential geometry, Springer-Verlag, New York, 1972.
3. J. L. Koszul, Lectures on transformation groups, Tata Institute Lectures on Mathematics and Physics, no. 20, 1965.
4. C. C. Moore, Amenable subgroups of semisimple groups and proximal flows, Israel J. Math., 34(1979), 121-138.
5. R. Phelps, Lecture on Choquet's Theorem, van Nostrand, Princeton, 1966.
6. V. S. Varadarajan, Groups of automorphisms of Borel spaces, Trans. Amer. Math. Soc. (1963), 191-220.
7. R. J. Zimmer, Induced and amenable actions of Lie groups, Ann. Sci. Ec. Norm. Sup., 11(1978), 407-428.
8. R. J. Zimmer, Ergodic Theory and Semisimple Groups, Birkhauser-Boston, 1984.
9. R. J. Zimmer, Semisimple automorphism groups of G-structures, J. Diff. Geom., 19(1984), 117-123.
10. R. J. Zimmer, Volume preserving actions of lattices in semisimple groups on compact manifolds, Publ. Math. I.H.E.S., 59(1984), 5-33.
11. R. J. Zimmer, Actions of lattices in semisimple groups preserving a G-structure of finite type, Erg. Th. Dym. Sys., to appear.
12. R. J. Zimmer, Kazhdan groups acting on compact manifolds, Invent. Math., 75(1984), 425-436.
13. R. J. Zimmer, On discrete subgroups of Lie groups and elliptic geometric structures, Rev. Math. IberoAmer., vol.1, to appear.
14. R. J. Zimmer, Lattices in semisimple groups and distal geometric structures, Invent. Math., to appear.
15. R. J. Zimmer, On the automorphism group of a compact Lorentz manifold and other geometric manifolds, preprint.